T0134205

OXFORD BIOLOGY PRIMERS

Discover more in the series at
www.oxfordtextbooks.co.uk/obp

Published in partnership with the Royal Society of Biology

HORMONES

HORMONES

Joy Hinson and Peter Raven

Edited by Ann Fullick
Editorial board: Ian Harvey, Gill Hickman, Sue Howarth

OXFORD
UNIVERSITY PRESS

Great Clarendon Street, Oxford, OX2 6DP,
United Kingdom

Oxford University Press is a department of the University of Oxford.
It furthers the University's objective of excellence in research, scholarship,
and education by publishing worldwide. Oxford is a registered trade mark of
Oxford University Press in the UK and in certain other countries

Published in the United States of America by Oxford University Press
198 Madison Avenue, New York, NY 10016, United States of America

British Library Cataloguing in Publication Data

Data available

Library of Congress Control Number: 2019933123

ISBN 978-0-19-883282-9

Printed in Great Britain by
Bell & Bain Ltd., Glasgow

PREFACE

Welcome to the Oxford Biology Primers

There has never been a more exciting time to be a biologist. Not only do we understand more about the biological world than ever before, but we're using that understanding in ever more creative and valuable ways.

Our understanding of the way our genes work is being used to explore new ways to treat disease; our understanding of ecosystems is being used to explore more effective ways to protect the diversity of life on Earth; our understanding of plant science is being used to explore more sustainable ways to feed a growing human population.

The repeated use of the word *explore* here is no accident. The study of biology is, at heart, an exploration. We have written the Oxford Biology Primers to encourage you to explore biology for yourself—to find out more about what scientists at the cutting edge of the subject are researching, and the biological problems they're trying to solve.

Throughout the series, we use a range of features to help you see topics from different perspectives.

Scientific approach panels help you understand a little more about 'how we know what we know'—that is, the research that has been carried out to reveal our current understanding of the science described in the text, and the methods and approaches scientists have used when carrying out that research.

Case studies explore how a particular concept is relevant to our everyday life, or provide an intimate picture of one aspect of the science described.

The bigger picture panels help you think about some of the issues and challenges associated with the topic under discussion—for example, ethical considerations, or wider impacts on society.

More than anything, however, we hope this series will reveal to you, its readers, that biology is awe-inspiring, both in its variety and its intricacy, and will drive you forward to explore the subject further for yourself.

ABOUT THE AUTHORS

Joy Hinson PhD, DSc, FRSB, PFHEA.
Joy Hinson is an emeritus professor of Queen Mary University of London. She graduated from the University of London with a degree in comparative physiology, then conducted research into adrenal endocrinology for her PhD and pursued a career in research, publishing over one hundred research papers and numerous reviews and book chapters. In 2006, Joy was awarded the Doctorate of Science degree by the University of London, in recognition of the contribution she has made to our understanding of the function of the adrenal cortex. Joy has taught endocrinology and reproduction to both medical students and biologists for many years, being awarded the Draper prize for outstanding teaching in 2009. She is a Principal Fellow of the Higher Education Academy and a Fellow of the Royal Society of Biology, serving since 2015 on the Education committee. Joy has co-authored a book with Dr Peter Raven on *The Endocrine System* and has most recently written a book on goats, entitled *Goat*!

Peter Raven MBBS, BSc, PhD, MRCP, MRCPsych, FHEA.
Dr Peter Raven graduated from St Bartholomew's Medical School in 1985. During his time at Barts he gained a BSc in biochemistry, a PhD in adrenal endocrinology, and an MBBS degree. After training as a doctor he developed an interest in psychiatry and completed his psychiatry training in 1995, winning the Gold Medal and Gaskell Prize of the Royal College of Psychiatrists in 1994. Peter worked as an academic psychiatrist and Faculty Tutor at University College London until 2014. He has published extensively on both endocrinology and psychiatry, including co-authoring *The Endocrine System* with Professor Hinson. He has a long-standing interest in teaching, focussing on medical students and junior doctors, and has contributed to several textbooks.

ACKNOWLEDGEMENTS

As ever, thanks to our families and friends for keeping us sane and providing us with fresh ideas, especially: Michaela and James, Ben and Ashley, Ruth and Harri, and Giles and Ali. A big thank you to Harri for trying out our drafts during her Biology A* level (although we can't take all the credit for her 'A*'). Thanks also to colleagues for their input. Any mistakes are, obviously, our own. And thanks to Ann—we couldn't have a more friendly, better informed, or more supportive editor.

CONTENTS

ABBREVIATIONS

ACTH	Adrenocorticotropic hormone
ADH	Antidiuretic hormone
AMP	Adenosine monophosphate
AVP	Arginine vasopressin
ATP	Adenosine tri-phosphate
Barts	St Bartholomew's Hospital
BPA	Bisphenol A
CA	Corpus allatum
CAH	Congenital adrenal hyperplasia
CEH	Cholesterol ester hydrolase
CJD	Creutzfeld Jacob disease
CRH	Corticotropin releasing hormone
DES	Diethylstilboestrol
DHEA	Dehydroepiandrosterone
DHT	Dihydrotestosterone
ELISA	Enzyme-linked immunosorbent assay
ENaC	Epithelial sodium channel
EPO	Erythropoietin
ER	Endoplasmic reticulum
FSH	Follicle stimulating hormone
GH	Growth hormone
GHRH	Growth hormone releasing hormone
GLP	Glucagon-like peptide
GnRH	Gonadotropin releasing hormone
hCG	Human chorionic gonadotropin
HRE	Hormone response element
HRT	Hormone replacement therapy
LH	Luteinising hormone
MR	Mineralocorticoid receptor
MSH	Melanocyte stimulating hormone
NGFβ	Nerve growth factor beta
NIH	National Institute of Health (USA)
NSAIDS	Non-steroidal anti-inflammatory drugs
NSC	Neurosecretory cells
ob gene	Obesity gene
OIF	Ovulation inducing factor
OXT	Oxytocin
POMC	Pro-opio-melano-cortin
PTG	Prothoracic gland
PVN	Paraventricular nucleus
RMET	The Reading the Mind in the Eyes test
SERM	Selective oestrogen receptor modifier
SHBG	Sex hormone binding globulin
SON	Supraoptic nucleus
SRY	Sex-determining region Y

T3	Tri-iodothyronine
T4	Thyroxine
THBG	Thyroid hormone binding globulin
TR	Thyroid hormone receptor
TRH	Thyrotropin releasing hormone
TSH	Thyroid stimulating hormone
TUE	Therapeutic use exemption
VT	Vasotocin
WADA	World Anti-Doping Authority

1 INTRODUCING HORMONES

Hormones are amazing. They control an extraordinary range of processes in our bodies and in the bodies of all complex animals and plants (see Figure 1.1). We often think of hormones as being slow-acting signalling molecules, just concerned with homeostasis, but the reality is much more interesting.

Figure 1.1 We made this wordcloud using the chapters in this book—how many of the words are familiar to you?

Wordclouds.com

Hormones can influence our appearance, our mood, when we eat and drink, how well we cope with stress, the health of our immune system, our relationships with other people, and every aspect of reproduction.

This book will look at hormones and disease, the impact of hormones on development, the role of hormones in sex and reproduction, how behaviour can be influenced by hormones, and finally, the use of hormones as drugs. We will mostly look at human biology and medicine, but explore some of the really interesting examples from other animals as well. For example, did you know that **vasopressin**—sometimes known as **antidiuretic hormone or ADH**—which acts on our kidneys to produce concentrated urine and so conserve water, makes some male fishes very aggressive?

It's helpful to start with a few key ideas about hormones:

Fact 1. Hormones are signalling molecules that travel in the blood between the cell that produces them and their target cell.

Fact 2. Hormones act by binding to, and activating, specific receptors which are found on target cells.

These two facts define hormones.

Fact 3. The study of hormones is called **endocrinology**. The **endocrine system** is the collection of all those parts of the body that either make or respond to hormones.

1.1   Hormones—fact or myth?

What hormones are and how they work were unknowns before the twentieth century. The functions of several endocrine glands were known, of course, because of the problems caused by damage to, or loss of the gland, but nobody knew that these glands produced chemical messengers. The first hormone was identified in 1902 by Ernest Starling, a doctor, and William Bayliss, a physiologist, working at University College London. They were studying the control of the exocrine pancreas, measuring the release of digestive juices in response to food entering the small intestine. They had thought that the pancreas was controlled by the nervous system, so they carried out an experiment where they cut the nerves to the pancreas, expecting that this would stop it from working. When they found that the pancreas carried on functioning, the theory of nervous control was disproved. Instead they found that the cells of the small intestine produced a chemical messenger, which travelled in the bloodstream to the pancreas and caused the release of digestive juices. They named this messenger *secretin* and recognized that they had discovered a completely new way of signalling in the body, which did not rely on the nervous system. In 1905, Ernest Starling was the first person to use the term *hormone* to describe secretin.

The chances are that you will never have heard of secretin. It isn't, frankly, a very interesting hormone and most textbooks of endocrinology hardly mention it. In general, hormones are studied when there is a possibility that they might be important in understanding disease. At the end of the twentieth century it was suggested that secretin may be involved in a behavioural disorder, but this was quickly disproved and secretin once again returned to being a note in the history books. The scientific discovery that is the easiest to make is not necessarily the most important!

In the intervening years since Starling discovered secretin we have learned a great deal about hormones. However, there are many myths about these astounding chemicals, and you may well be familiar with some of them. So before we look at what we know about hormones, we will visit some of the most widely held misconceptions and debunk them.

1. **Hormones are slow acting.** Some are—but some really aren't. Just think about **adrenaline**. If you were standing in a field with a bull running towards you (see Figure 1.2), the 'fight-or-flight' reaction would need to be pretty quick if you wanted to stay alive. You couldn't wait for a slow homeostatic kind of response over hours or days, or you'd be trampled by the bull. Adrenaline has a really fast action, so when it is released it helps you to quickly jump over the fence and out of the way.

But it is true that the endocrine system generally operates at a slower pace than the nervous system and most hormones don't have such a fast action.

2. **The endocrine system is completely separate from the nervous system.** The standard way of describing the nervous system and the endocrine system is to treat them as completely different things. But

Figure 1.2 When your body recognizes a threat such as this bull, it will release adrenaline to have an immediate effect—whether or not the threat is real

doru dumitru/Shutterstock.com

the truth is that there is a lot of overlap and interplay between the endocrine system and the nervous system. Think back to adrenaline for a moment because this is a really good example. It is released from nerve endings so is part of the nervous system, where it functions as a neurotransmitter in the sympathetic nervous system. But instead of just being released into the synaptic cleft, it is also released into the blood, where it travels to different tissues and acts by binding to receptors. As we have just seen, this is the definition of a hormone. This mixture of types of action makes adrenaline a **neurohormone**: a hormone that is released from nerve endings into blood (see Fig. 1.3). Other neurohormones include vasopressin (antidiuretic hormone or ADH) and **oxytocin**. We'll look at those later on.

There are also many examples of hormones that act on the nervous system, especially in the brain. Hormones influence many aspects of behaviour by acting on different parts of the brain. We'll look at some examples of this in Chapter 4. The endocrine and nervous systems are clearly not as separate as we might think.

Figure 1.3 This diagram shows the structural difference between the nervous system and the endocrine system

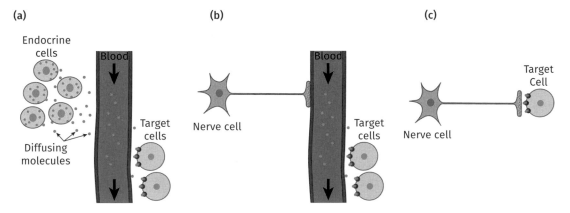

(a)

Endocrine cells

Blood

Diffusing molecules

Target cells

(b)

Nerve cell

Blood

Target cells

(c)

Nerve cell

Target Cell

(a) shows hormones being released into blood from a hormone-producing cell and travelling around the body to interact with their target tissue. Panel (b) illustrates neuroendocrine signalling. The hormone is released from nerve terminals, like a neurotransmitter, but travels in blood like a hormone. In panel (c) you can see that the chemical messenger (neurotransmitter) is delivered by neurons directly to the target tissue.

Reproduced from Pocock et al, *Human Physiology* 5/e. By permission of Oxford University Press

3. **Hormones are all released by endocrine glands**. We have a very clear idea of what the 'endocrine system' looks like, in terms of the glands that produce hormones, from the typical textbook diagram shown in Figure 1.4.

However, this is a real over-simplification. Hormones are released by many different cell types, not just the major endocrine glands. Going back again to adrenaline, for example, most of the adrenaline in a 'fight-or-flight' response comes from the nervous system and relatively little from the adrenal glands. If somebody has their adrenal glands removed they still have a

Figure 1.4 The more widely recognized endocrine organs of the body

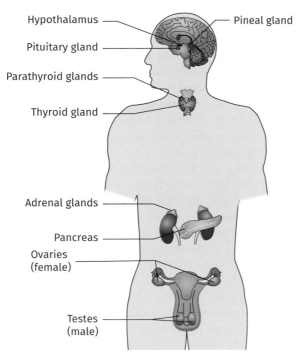

Hypothalamus ——
Pituitary gland ——
Parathyroid glands ——
Thyroid gland ——
—— Pineal gland

Adrenal glands ——
Pancreas ——
Ovaries
(female) ——
Testes
(male) ——

'fight-or-flight' response and no adrenaline replacement therapy is needed (although the adrenal gland makes other important hormones which do need to be replaced).

If you were asked to name the largest endocrine organ in the body you might wonder whether the pancreas or the thyroid would be the largest. In fact, neither of these is anything like the largest hormone-producing tissue in the body and the right answer would be the skin. It doesn't appear on any diagram showing the endocrine organs, but the skin is very important indeed in the production of the hormone vitamin D.

The other very large endocrine tissue in the body is adipose tissue: the collection of fat cells that is present just beneath the skin and in the abdomen, surrounding the kidneys and other internal organs. Fat cells produce a hormone, called **leptin**, that acts on the brain to control how hungry we feel.

4. **Every hormone has a target tissue**. This is another oversimplification. Before molecular techniques were used we had to identify the target tissues of hormones by looking for a physiological action. Some hormones appear to have relatively simple actions. A good example is vasopressin, which acts on the kidney to increase water reabsorption. It does this by activating vasopressin receptors in the kidney nephron which causes water channels to open, allowing water to leave the urine and return to the blood. You might think that this was quite enough, but vasopressin has another major action on a completely different target tissue.

The effects of vasopressin were established by injecting the hormone into experimental animals and observing the result. The main effect of vasopressin was to cause a large increase in the animals' blood pressure. Part of the observed increase in blood pressure could be explained by the actions of vasopressin on the kidney. This is because increased water reabsorption results in an increased blood volume, which in turn increases blood pressure. But not all of the blood pressure effect could be explained in this way and scientists realized that vasopressin was also acting directly on blood vessels. The hormone was causing blood vessels to constrict and this was causing much of the rise in blood pressure. We now know that this is a major effect of vasopressin and so this hormone has two major target tissues: the kidney and blood vessels.

One more inaccuracy you may well have heard in your education is that this hormone is called 'antidiuretic hormone' or ADH. In fact scientists and doctors call this hormone arginine vasopressin (AVP), or just vasopressin, not antidiuretic hormone because we now know it has so many other actions. It is now only called ADH in school books and some of the simpler (and more out-of-date) textbooks. Arginine vasopressin is a member of a family of related hormones with a really interesting range of effects in different types of animals. Even then, the action of AVP is relatively straightforward compared to a hormone like thyroxine.

Thyroxine, produced by the thyroid gland, is a good example of a hormone that does not have a single target tissue. Classical experiments in the nineteenth century involving removal of the thyroid gland from dogs showed that the dogs died, although it wasn't obvious why death occurred. The deaths could be prevented by treating the dogs with extract of thyroid gland. In people it was known that removal of the thyroid gland caused a condition called myxoedema: a swelling of the skin and tongue, with hair thinning, constipation, weight gain, and general mental and physical slowing down. It is clear that thyroid hormones don't just affect one single tissue type.

Molecular techniques have shown that there are several different genes which code for thyroid hormone receptors (TR) and these genes are expressed in almost all tissues of the body. We know that thyroid hormones have a wide range of effects throughout the body, affecting basal metabolic rate, brain function, the heart and blood vessels, growth, and just about every other aspect of physiology. Without thyroid hormone, we die.

5. **Thyroxine and testosterone are hormones.** Actually, some of the 'hormones' you have heard of aren't really hormones. For example, both testosterone and thyroxine are hormone **precursors**, or **pre-hormones**. In order to bind to their receptors these pre-hormones need to be converted by enzymes into their active form.

Thyroxine is always described as the main hormone produced by the thyroid gland. That statement is true to a very limited extent, in that thyroxine is produced in the largest quantities and we now know that thyroxine does have some direct hormonal actions. However, most of the effects that are given as the classical actions of thyroxine, such as its effect on metabolism, are actually actions of T3 or **tri-iodothyronine**, not thyroxine. Before binding to a receptor, thyroxine is metabolized by enzymes in the target tissues and converted into T3. It is the T3 that then binds to the receptor and has the hormonal effect. T3 really is a hormone, because it is also produced by the thyroid gland, in smaller amounts than thyroxine, and circulates in blood.

Testosterone is another good example of a pre-hormone. We know that, as the main male sex hormone, testosterone increases sperm production in the testes. It also causes men to have deeper voices and bigger muscles than women. Testosterone is made in the testes. When it acts to increase sperm production it does this by diffusing through the intercellular spaces in the testes. It doesn't need to travel in blood to have its effect, so it isn't acting as a hormone to have its effect on sperm production. When testosterone acts on the vocal cords or on muscle it has to travel from the testes to these target tissues, but it gets converted to dihydrotestosterone (DHT) on the way. It is DHT that binds to the receptor in the muscle or vocal cords and has the effect of increasing strength and deepening the voice. In the testes, testosterone is acting as a local regulatory factor; elsewhere in the body it acts as a pre-hormone. However, *every* book describes testosterone as the principal androgenic hormone and we shall follow that convention.

6. **Vitamins and hormones are completely different things.** Not all vitamins are really vitamins. Vitamin D really isn't a vitamin because most of the vitamin D we need to be healthy is made in the body and not obtained in our diet. It is actually quite difficult to get enough vitamin D from our diet to stay healthy. The only foods that contain significant amounts of vitamin D are oily fish and most of us don't eat enough oily fish to be able to rely on this as a source of vitamin D (see Figure 1.5). In fact, like testosterone and thyroxine, vitamin D is a pre-hormone. It is mostly made in the skin by the action of sunlight.

Figure 1.5 You would need to eat a lot of oily fish to get as much vitamin D as you could get from the effects of sunshine on your skin. If you have darker skin you need more sunshine to produce enough vitamin D. Sunscreen stops the production of vitamin D.

Davdeka/Shutterstock.com

The vitamin D is carried around the body in the bloodstream. But before binding to its receptor, vitamin D needs to be activated by two enzymes, one in the liver and the second in the kidney. Once it is activated it is able to bind to its receptor in bone and other tissues. The activated form of vitamin D is called **calcitriol** (or 1,25 dihydroxycholecalciferol or vitamin D3) and is very important for healthy bones in children and adults. Vitamin D is so important that, in many countries, it is added to everyday foods to try to ensure that everybody has enough. These fortified foods include milk, orange juice, and breakfast cereals.

7. **Hormones are one particular type of chemical.** This is really not true. Hormones come in all sorts of forms. They all share the common properties of travelling in blood and binding to receptors, but *chemically* they are very different indeed. Some are peptides, some are steroids, and some come from the amino acid, tyrosine. These chemical differences affect the way that they act and the types of receptors that they activate. Now we have debunked some of the common hormone myths, let's start to look at these amazing compounds in more detail.

Scientific approach 1.1
Berthold's chickens: the first endocrine experiment

Classical experimental approaches to endocrinology start by asking the simple questions. The first question is 'What happens if we take the hormone away?' The easiest way to do this was to surgically remove the endocrine gland and observe what happened. There is a very simple demonstration of how this works. It has been known for hundreds of years that removing the testes from male animals makes them less aggressive. Since ancient times male farmyard animals have been castrated (had their testes removed) to make them less aggressive and easier for a farmer to handle. From these simple experiments we can conclude that the testes affect the brain of male animals.

The second simple question in endocrinology is 'What happens if we put back the hormone (or endocrine tissue) that we removed?'

The first person to conduct this type of experiment was a German physiologist called Arnold Berthold. In 1849 Berthold was investigating the link between the testes and the brain. His **hypothesis** was that there was a neural link between the two tissues which allowed the testes to affect the brain. He worked with young male chickens which, as they mature into adult cockerels, grow a bright red coxcomb on their heads and start to display aggressive behaviour typical of cockerels: crowing, strutting, chasing hens, and fighting other cockerels. When cockerels are castrated they don't grow a coxcomb and they don't behave like adult cockerels. Berthold took eight immature cockerels and castrated six of them. He left two intact as 'controls' to show what happens without any intervention. He then took two of the castrated cockerels and transplanted one of their own testes back into them. He took two more and transplanted a different cockerel's testes into them. He then waited for them to become fully grown.

The results of his experiment are shown in Figure A. As you can see in Figure A, the control group developed into normal adult cockerels, with full coxcombs and showed typical cockerel behaviour. The castrated group didn't develop coxcombs and didn't behave like cockerels. Both the transplanted groups, even though they only received one testis, developed into normal adult cockerels with coxcombs and cockerel behaviour.

Figure A This diagram shows the results of Berthold's experiments on cockerels. As long as the young birds were exposed to testosterone they developed male characteristics. The evidence from these experiments allowed Berthold to hypothesize hormones as blood-borne factors.

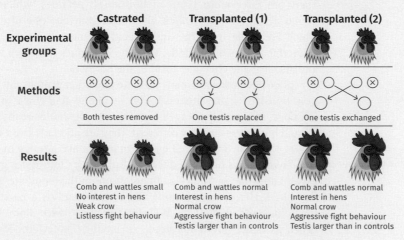

Berthold killed the transplanted chickens and examined the transplanted testes. He saw that there were no nerves supplying the transplanted testes, but a blood supply had developed. He concluded that a blood-borne factor must be produced by the testes which caused the coxcomb to grow and affected the brain to produce typical cockerel behaviour. He also noticed that the transplanted testis was larger than either of the testes of the control group of cockerels. Why do you think that might have happened?

Berthold's experiments led to the discovery of the endocrine system. We still carry out similar experiments today, removing part of the endocrine system and putting it back again. Today we mostly use molecular techniques where a gene can be knocked out from a specific cell type at a particular stage of development. Using sophisticated molecular methods to silence or amplify genes is the modern equivalent of the classical take out/put back endocrine experiments.

> **? Pause for thought**
>
> There is another interesting aspect of Berthold's experiment that was way ahead of its time. Berthold's hypothesis was that there was a neural link between the testes and the brain. The experiments he carried out were designed to *disprove* this hypothesis, because it is always easier and more reliable to disprove a hypothesis than to try to prove it. Why do you think this is?
>
> If Berthold had set out to try to *prove* his hypothesis, how might he have approached this experimentally?

1.2   Types of hormones and how they are made

There are three main types of hormones: peptides, which are chains of amino acids; steroids, which are made from cholesterol; and hormones, which are made from the amino acid tyrosine. Table 1.1 lists the major endocrine tissues of the body, with the hormones that each tissue produces, and also lists the chemical form of the hormone.

Peptide hormones are made just like any other protein: by gene transcription, translation, and processing of the amino acid chain to make the final hormone. It is then packaged into secretory granules ready to be released. Steroid hormones are very different: they are all made from **cholesterol** (see Figure 1.6), which is altered by enzymes to make a range of hormones with similar structures but very different functions. Cholesterol gets a bad press: we all know that high levels of cholesterol in the blood are bad news, causing atherosclerosis and leading to heart disease. As a result, many people try to limit the amount of cholesterol they eat, with many foods being labelled as either 'low cholesterol' or having 'cholesterol-lowering' effects. Fortunately, in humans, the cells that make steroid hormones are able to produce their own cholesterol in addition to using cholesterol from the diet, so there is always enough cholesterol to make the steroid hormones.

Figure 1.6 Cholesterol is the starting point for the production of all steroid hormones

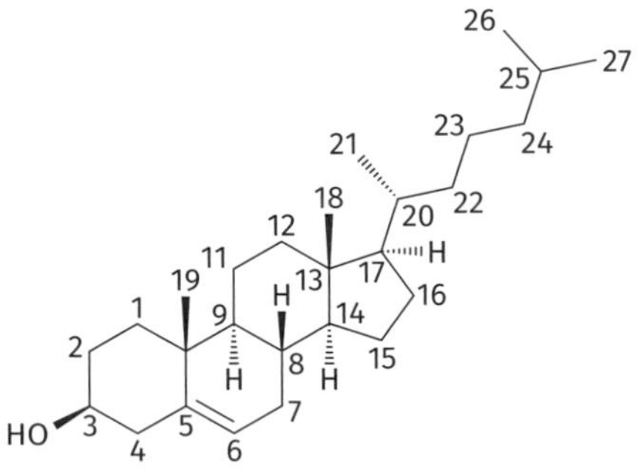

Table 1.1 The major endocrine tissues of the body

Gland	Hormone	Type of hormone
Hypothalamus	Corticotropin releasing hormone (CRH)	Peptide
	Gonadotropin releasing hormone (GnRH)	Peptide
	Growth hormone releasing hormone (GHRH)	Peptide
	Somatostatin	Peptide
	Thyrotropin releasing hormone (TRH)	Peptide
	Dopamine	Modified amino acid
Pituitary	Adrenocorticotropin (ACTH)	Peptide
	Follicle stimulating hormone (FSH)	Glycoprotein
	Growth hormone (GH)	Peptide
	Luteinizing hormone (LH)	Glycoprotein
	Prolactin	Peptide
	Thyroid stimulating hormone	Glycoprotein
	Oxytocin	Peptide
	Vasopressin	Peptide
Thyroid	Thyroxine (T4)	Modified amino acid
	Tri-iodothyronine (T3)	Modified amino acid
	Calcitonin	Peptide
Parathyroid	Parathyroid hormone	Peptide
Adrenal	Aldosterone	Steroid
	Cortisol	Steroid
	DHEA	Steroid
	Adrenaline (epinephrine)	Modified amino acid
	Noradrenaline (norepinephrine)	Modified amino acid
Pancreas	Insulin	Peptide
	Glucagon	Peptide
Stomach and gut	Gastrin	Peptide
	Ghrelin	Peptide
	Many other peptides	Peptide
Ovaries	Oestradiol	Steroid
	Progesterone	Steroid
Testes	Testosterone	Steroid
Kidneys	Erythropoietin (EPO)	Peptide
	Activated vitamin D3	Modified steroid
Adipose tissue	Leptin	Peptide
Pineal gland	Melatonin	Modified amino acid

It is helpful to go through some of the words used to describe hormone production. The process of making hormones of any type is called **hormone synthesis**. Sometimes, but not always, there is a difference between hormone synthesis and release, called **secretion**. Some hormones, particularly peptide hormones, are actively secreted. Others, such as steroids, are released from the cell as they are made. The process of steroid hormone synthesis is called **steroidogenesis**.

Peptide hormones are the most numerous type of hormone in the body (see Figure 1.7). They range in size from the tiny peptides, made up of fewer than ten amino acids, to huge glycoproteins with a molecular mass of between 20,000 and 30,000 Daltons. These are large polypeptides with sugar **residues** attached. The smallest peptide hormones are usually just a single chain of amino acids, but the larger ones may consist of two chains joined by disulphide bridges, or a long single chain, stabilized by disulphide bridges.

Insulin is a polypeptide hormone produced by the beta cells of the islets of Langerhans in the pancreas. It consists of two polypeptide chains which are the product of a single gene. A **precursor peptide** is produced, which goes through post-translational processing to put the disulphide bridges in place and to remove the amino acid sequence that originally linked the two chains. The insulin is packaged into a **secretory granule**, which stays within the cells, forming a reservoir of hormone ready for release. Peptides are hydrophilic and generally quite large, so they do not just diffuse out of the cell, but need to be actively secreted. The insulin is secreted when the secretory granule fuses with the plasma membrane of the beta cell. It does this in response to a specific signal; in this case, a rise in blood glucose.

Steroid hormones are made very differently. These hormones are made mainly by the ovaries, testes, and the adrenal cortex. Cells that produce steroids also contain lipid droplets. These are a pool of cholesterol, stored as cholesterol esters, which is ready to be used in steroidogenesis. When the cell receives a signal that steroid hormones are needed, cholesterol from the pool travels first to the **mitochondria** and then to the **endoplasmic reticulum**. These organelles are the places where the enzymes that make the steroid hormones are located. Steroidogenesis has a common pathway at the start, so the first steps of hormone production are the same in all the tissues that make steroids. However, the adrenal gland contains different enzymes to the testes and ovaries and so produces a different set of hormones (see Figure 1.8).

In contrast with peptide hormones, steroid hormones are small and lipophilic, and they can easily cross the plasma membrane of cells. There is no store of pre-made hormone in steroid-secreting cells and these hormones are released from the cell, probably by simple diffusion across the plasma membrane, as soon as they are made.

The hormones made from the amino acid tyrosine are the thyroid hormones and the **catecholamines**, adrenaline, **noradrenaline**, and **dopamine** (see Figure 1.9). These hormones are made in very different ways and have quite different properties. The thyroid hormones, thyroxine (T4) and tri-iodothyronine (T3), consist of two tyrosine residues joined together, with iodine attached to each tyrosine. Thyroxine has four iodine atoms and tri-iodothyronine has three. Despite the molecular mass of these

Figure 1.7 Insulin is a medium-sized polypeptide hormone which is made from two polypeptide chains joined by disulphide bridges, while growth hormone is a single polypeptide chain stabilized by disulphide bridges

Insulin

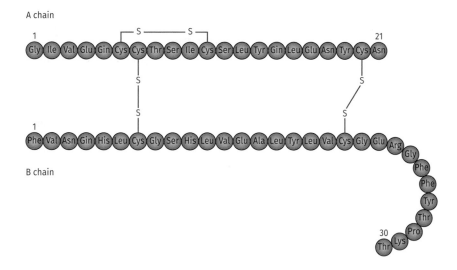

Growth hormone

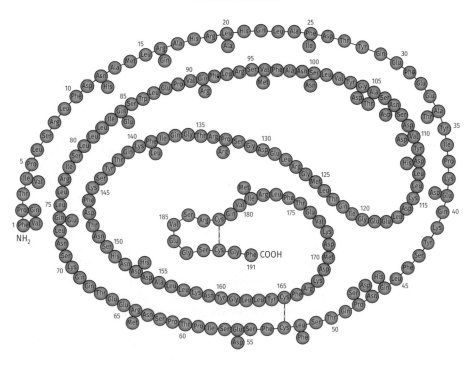

Figure 1.8 The pathway of steroid hormone production (steroidogenesis) starts the same way in all the steroid-producing tissues, but then diverges, depending on the different types of enzyme present.

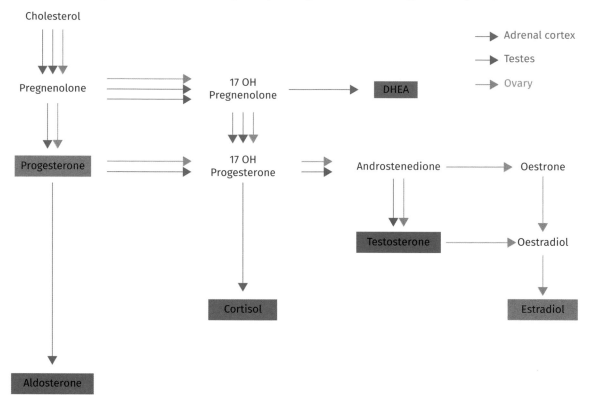

hormones being about twice the size of the steroid hormones, they are also able to cross the plasma membrane of a cell by simple diffusion. They are made as part of a very large protein which is stored in a series of pools in the thyroid gland, and processed in the thyroid cells in response to thyroid stimulating hormone (TSH). A diagram showing this process in more detail is given in Chapter 2.

The other hormones made from tyrosine, the catecholamines, are synthesized like steroids but then behave much more like peptides. Single tyrosine residues go through a series of reactions, catalysed by specific enzymes in the cells of the adrenal medulla. This metabolism of tyrosine produces the hormones adrenaline and noradrenaline, which are sometimes called **epinephrine** and **norepinephrine**. Once the hormones are formed they are stored in secretory granules in the cells of the adrenal medulla until the signal arrives for their release. The adrenal medulla is part of the autonomic nervous system and so is directly supplied with nerves. More specifically it is part of the sympathetic system and so regulates 'fight-or-flight' responses. When somebody senses a threat that needs an immediate response, action potentials to the adrenal medulla cause depolarization of the adrenal

Figure 1.9 The amino acid tyrosine is the starting point for the production of both thyroid hormones and catecholamines such as adrenaline

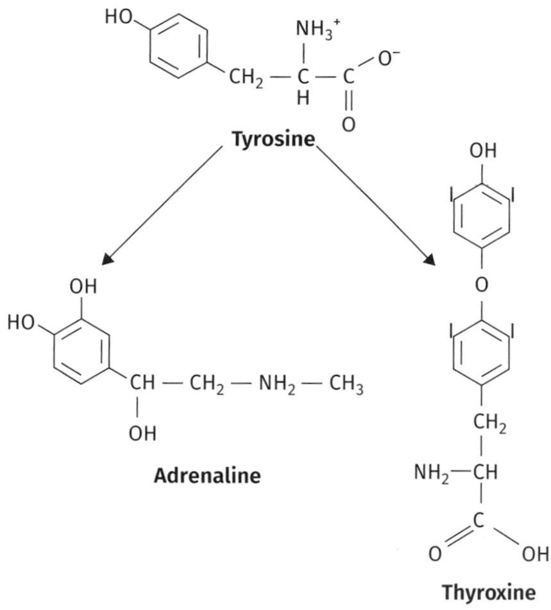

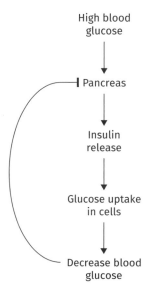

Figure 1.10 This flow diagram illustrates the negative feedback control of insulin production. We have used the convention that an arrow indicates activation and a line with a bar across shows a block or inhibition. So high blood glucose activates the pancreas to produce insulin, which results in a decreased blood glucose, which stops the pancreas from producing more insulin.

medullary cells. The action potentials cause the release of adrenaline and no-radrenaline from the medullary cells, in the same way as neurotransmitters are released from nerve terminals. In the adrenal medulla the catecholamines are released into blood, not the synaptic cleft, and so are hormones rather than neurotransmitters in this situation (see Figure 1.3).

1.3   Control of hormone production

Most hormones are regulated by **negative feedback control** (see Figure 1.10). Insulin is the most straightforward example of this. An increase in blood glucose concentration activates specific glucose receptors on the beta cells of the endocrine pancreas, which signal the cell to release insulin. Insulin acts to reduce blood glucose concentration. This means that the glucose receptor is no longer activated and the beta cell stops releasing insulin. So insulin release is controlled by negative feedback effects of blood glucose. In this way glucose concentrations in blood do not become too low. Because a low blood sugar is much more immediately dangerous to the body than a high blood sugar, there is a whole group of hormones whose actions stop blood glucose from falling too low: the decrease in insulin release is just one part of this mechanism.

Some hormones are released as part of a hormone cascade (see Figure 1.11). The processes of peptide hormone release and steroid hormone production are both controlled by biological cascades. As with other

Figure 1.11 A hormone cascade, which typically involves the hypothalamus producing a hormone which acts on the pituitary to produce a second hormone. This acts on a target tissue to produce a third hormone. This third hormone has effects on the body, but also has a negative feedback effect on both the hypothalamus and pituitary, stopping further activation of the cascade.

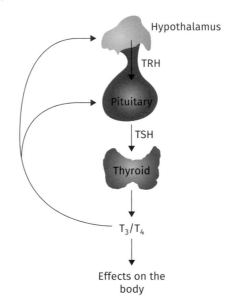

The control of thyroid hormone production is a classic example of both a hormone cascade and the effect of negative feedback.

biological cascades, this sequence of hormone activations allows a large and rapid response to altered needs and a very fine control as a result of multiple negative feedback points.

Thyroid hormones are a good example of this. Blood levels of thyroid hormone are usually maintained at a set level. But the set level needs to be changeable, as a number of different body states or environments cause an increase in the thyroid hormone requirement. These include exercise, exposure to cold, and pregnancy. Any one of these causes an activation of the hypothalamus, a part of the base of the brain. The hypothalamus releases a hormone which activates the pituitary gland. This hormone is called **thyrotropin releasing hormone (TRH)** and is a tiny peptide hormone, of only three amino acids. TRH stimulates the release of **thyroid stimulating hormone (TSH)**, one of the very largest glycoprotein hormones, from the pituitary gland. This binds to TSH receptors on the thyroid cells, activating the cells, and causing the production of more thyroid hormone (T4 and T3). When blood levels of thyroid hormone have increased to the new set level, the thyroid hormone acts on receptors in the hypothalamus and pituitary. As a result, the hypothalamus decreases the amount of TRH released and the pituitary decreases the amount of TSH released, both of which lead to decreased output of TSH and so the thyroid is no longer activated. This is the negative feedback effect of thyroid hormone, acting to shut down the pathway that caused it to be produced, once levels are high enough. The new set level of thyroid hormone is maintained by small changes in activation of this cascade.

1.4   Hormones in the blood

It is one of the defining features of hormones that they are all transported in blood from the cells that make them, to their target cells, where they have their effects. Let's take a look at how hormones are transported in blood and what happens to them there.

The chemical nature of hormones influences how well they dissolve in water. As blood plasma is mainly water, the solubility of different hormones in water determines how well they dissolve in blood. Peptides and catecholamines are very soluble in water, despite the very large size of some polypeptides. They are usually carried in blood in solution. Steroids and thyroid hormones are much less soluble in water, although they are small molecules. When these hormones are released into blood they bind loosely to plasma proteins, such as albumin. Several of these hormones also have their own specific binding proteins which aid the solubility of the small hormones by transporting them through the blood. Thyroid hormones, for example, bind to **thyroid hormone binding globulin (THBG)** and both androgens and **oestrogens** bind to **sex hormone binding globulin (SHBG)**. These binding proteins carry about 95% of their specific hormone in the blood and the rest is either dissolved in the blood plasma or loosely bound to albumin.

While a steroid is bound to a large plasma protein it can't bind to a receptor and so the bound hormone is not **biologically active**. It is only the 5% of the hormone that is not bound to a plasma protein that is available to have a biological effect. As a result, when you need to measure hormone concentrations in blood, it is important to know how much of what you are measuring is unbound and how much is bound to transport proteins. In a way these binding proteins act to provide a mobile store of hormone in the blood. They also act to increase the **half-life** of a hormone in blood. Why do you think that is?

As hormones travel around the body in the blood, they are metabolized and degraded by enzymes in the cells and tissues that they pass through. Peptidase enzymes on endothelial cells in the blood vessels break down peptide hormones, while steroid hormones are metabolized mainly in the liver or excreted by the kidney in an unchanged form. Peptide hormones are also metabolized when they bind to their specific receptor on the target cell. The half-life of a hormone is a measure of the amount of time it takes for half the hormone to be broken down or excreted. It can be measured experimentally by injecting a small amount of a 'tagged' form of the hormone intravenously (typically it would be tagged with a radioactive label) and then seeing how quickly the tagged hormone disappears from the blood. Adrenaline and noradrenaline have the shortest half-lives of any hormones; it takes only a few seconds before they are metabolized. The thyroid hormone, thyroxine, has the longest half-life, at around six to seven days. Most peptide hormones have a half-life of several minutes, while steroids bound to their plasma binding proteins have a half-life of around ninety minutes.

1.5   Hormone receptors and mechanisms of hormone action

By this point you may have begun to notice a pattern. Large, water-soluble hormones, mainly peptides, need a specific mechanism for getting out of the cell that produces them, but they are quite soluble in water so don't need a transport mechanism in blood. Steroids and thyroid hormones, which are lipophilic, don't require a specific secretion mechanism, but need binding proteins to increase their solubility in the blood. These characteristics also influence the receptors that different types of hormones interact with.

Hormone receptors are all large proteins. They have a region that binds the hormone and another region that works to relay the signal to influence the working of the cell. Hormone receptors can be found in the plasma membrane, in the cytoplasm, and in the nucleus of a cell. Generally, the chemical characteristics of the hormone determine which type of receptor it binds to.

Peptides and catecholamines

Peptide hormones and catecholamines all have their receptors located in the plasma membrane of the cell, with the hormone-binding site facing the extracellular space. This means that the hormones do not need to enter the cell in order to bind to the receptor. However, this does also mean that there has to be a mechanism for getting the signal into the cell. The receptors for peptide hormones have different ways of passing the signal into the cell, but all of these result in the same end point: the phosphorylation of target proteins. Phosphorylation describes the process of adding a phosphate (PO_4^{3-}) group to a biological molecule. This can change the activity of a protein, particularly if the protein is an enzyme. Some enzymes only become active once they have been phosphorylated. By causing the phosphorylation of enzymes inside a cell, the enzymes become activated and the activity of the cell changes.

When a peptide hormone binds to its cell membrane receptor the receptor changes shape. This simple change activates an enzyme called a **protein kinase**, which phosphorylates specific proteins inside the cell. Sometimes the protein kinase is a part of the receptor protein and sometimes it is a separate protein which is recruited when the hormone binds. Activation of a single protein kinase can result in phosphorylation of another, and then another kinase. This is called a protein kinase cascade and usually leads to changes in gene expression, activating the cell cycle, for example.

For many peptide hormones, **cyclic AMP** (cAMP or adenosine monophosphate) acts as a **second messenger**, as you can see in Figure 1.12. Cyclic AMP activates a protein kinase by binding to it and changing its shape. However,

Figure 1.12 The binding of peptide hormones to receptors on the surface of a cell often triggers the formation of a secondary messenger, such as cAMP. It is this secondary messenger that triggers responses in the cell.

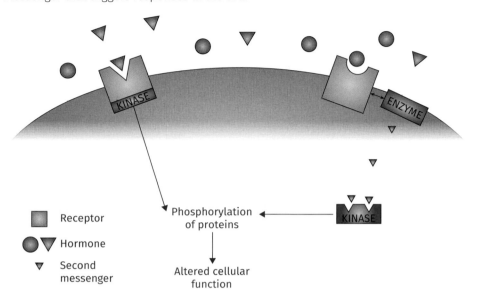

other peptide hormones, such as insulin and growth hormone, have receptors which are directly linked to a protein kinase, without the need for a second messenger system. Ultimately, all peptide hormones have their effects by binding to cell-surface receptors and activating kinases. You can see an example in Case study 1.1.

Steroids and thyroid hormones

Thyroid hormones and steroids, unlike peptide hormones, can freely cross the plasma membrane and enter the cell. Their receptors are located in the cytoplasm and nucleus of the cell. A hormone binding to its intracellular receptor may seem to be at the end of its signalling pathway, but in order for the hormone to actually affect the function of the cell, the binding of the hormone needs to trigger a response in the target cell.

The receptors for thyroid hormones and steroids are all hormone regulated **transcription factors**. These alter the function of the cell by directly affecting gene transcription. When the hormone binds to one of these receptors, in the cytoplasm or nucleus of the cell, the hormone–receptor complex then binds to the promoter region of a DNA sequence. This is the part of a gene that is not transcribed, but which determines whether the gene itself is transcribed. By changing the rate of transcription, or expression, of certain genes, the cell produces a different set of proteins and the cell function is changed. Hormone receptors are all linked to **intracellular signalling pathways** that ultimately alter cell function (see Figure 1.13).

Figure 1.13 Steroid and thyroid hormones act to alter gene transcription in target cells

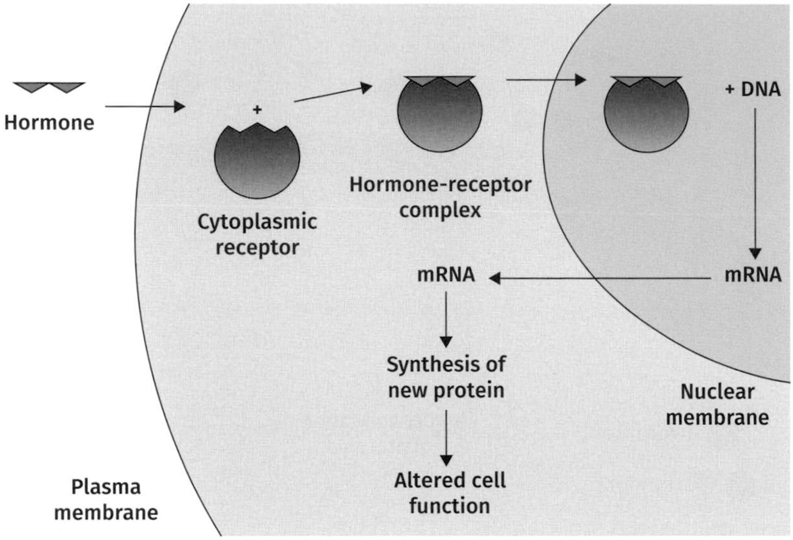

Reproduced from Pocock et al, *Human Physiology* 5/e. By permission of Oxford University Press

Steroid and thyroid hormones mostly have very complex effects on cells and it is difficult to pinpoint an example where a hormone affects only one gene and so affects the cellular function. For example, when androgens, the male sex steroids, act on muscle to increase muscle bulk and strength, this is the result of the activation of many different genes. There is one example, however, where a steroid hormone has a straightforward effect, mainly by affecting the expression of a single gene: the action of aldosterone in the kidney. You can read about this in Case study 1.1.

Case study 1.1

LH and testes, aldosterone and sodium balance

Many hormones have multiple different actions in the body. Here we will look in detail at the specific actions of individual hormones on particular cell types—the Leydig cells of the testes and the epithelial cells lining the kidney tubules.

LH acting on the testes

An example of a peptide hormone acting to change the function of a cell by causing phosphorylation of a protein is the effect of luteinizing hormone (LH) on testosterone production by the Leydig cell of the testis (see Figure A).

LH binds to its receptor, causing it to change shape. This releases a shuttle protein which moves through the plasma membrane and activates an

Figure A Actions of LH on the Leydig cell of the testis. Activation of the steroid synthesis pathway in the Leydig cell results in the production of testosterone.

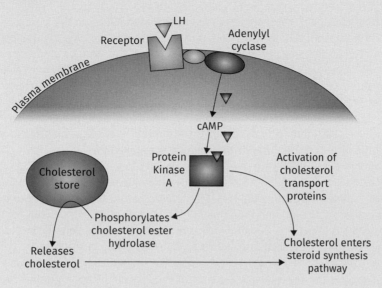

enzyme. The enzyme is called adenylyl cyclase and when it is activated it produces a signalling molecule called cyclic AMP (cAMP). Cyclic AMP belongs to a group of substances called 'second messengers' that are produced by receptor-mediated activation of an enzyme and which in turn activate a kinase within the cell. In the Leydig cell cyclic AMP binds to protein kinase A and activates it. This protein kinase phosphorylates several proteins in the cytoplasm of the Leydig cell, including an enzyme called cholesterol ester hydrolase (CEH). Phosphorylation activates the CEH and it moves towards the lipid droplets in the cytoplasm of the Leydig cell. As we have seen, these lipid droplets contain cholesterol esters, ready to be used in steroidogenesis. Before this can be used the cholesterol has to be de-esterified by CEH. The activated CEH acts on the cholesterol esters and releases free cholesterol into the cell, which is then used in the **synthesis** of steroid hormones, in this case testosterone. By using a second messenger and a protein kinase, LH can bind to a receptor on the cell surface and cause changes within the cell, without needing to enter the cell.

❓ Pause for thought

The synthesis of steroid hormones from cholesterol involves many steps, carried out by different enzymes. It would be wasteful if stimulatory hormones like LH had to activate each of these steps independently in order to trigger steroidogenesis. Instead, the whole steroidogenic apparatus is constantly ready for action apart from one critical control point, called the **rate-limiting step** of the pathway. It is the rate-limiting step that LH activates to regulate the whole pathway. Would you design things so the rate-limiting step is at the beginning or the end of the steroidogenic pathway?

What is the rate-limiting step in peptide hormone synthesis?

Aldosterone and the kidneys

When you think of a hormone that acts on the kidney, ninety-nine times out of a hundred you will think of vasopressin (ADH), which controls water up-take in the kidney. But there is another hormone that acts on the kidney and controls sodium balance. This hormone is a steroid called aldosterone (pro-nounced al-doh-steer-own). Aldosterone is one of the steroids made by the adrenal cortex. Its main action is to increase the uptake of sodium from the renal tubules, returning it to the blood. Aldosterone does this by binding to a specific receptor in the cells of the renal tubule. This receptor is called MR, or mineralocorticoid receptor. The receptor is found in the cytoplasm of the cell and when aldosterone binds to the MR the hormone–receptor complex moves to the nucleus. There it binds to a specific site, called the mineralocorticoid response element (MRE) in the promoter region of a gene. Altering the rate of gene transcription requires two sets of the hormone–receptor complex, which form a **dimer** and bind to the DNA. This binding causes other proteins, called co-activators, to bind and these initiate the process of gene transcription, or switching on the gene. Sometimes this is called increasing gene expression. The particular gene that is switched on by aldosterone in the kidney is the gene that encodes a protein called epithelial sodium channel (abbreviated to ENaC). As the gene is activated, the cell produces more copies of the ENaC protein. ENaC is trafficked to the plasma membrane of the kidney epithelial cell, facing the filtrate. It allows the epithelial cell to take up sodium from the urinary filtrate. This sodium is then pumped out of the cell and into the blood by the active transporter of sodium, the sodium–potassium ATPase enzyme, sometimes referred to as the 'sodium pump' (see Figure B).

Figure B This diagram shows the structure of a kidney and the arrangement of a single nephron. Aldosterone acts on epithelial cells in the distal collecting tubules. ENaC is a sodium channel protein and HRE is hormone response element: the section of DNA that binds the the hormone-receptor complex. Aldosterone increases the rate of transcription of the gene which codes for ENaC. This results in an increased number of sodium channels to reabsorb sodium from the urinary filtrate, down a concentration gradient. The sodium concentration inside the cell is kept low by the actions of Na^+/K^+ ATPase, which actively pumps sodium out of the cell.

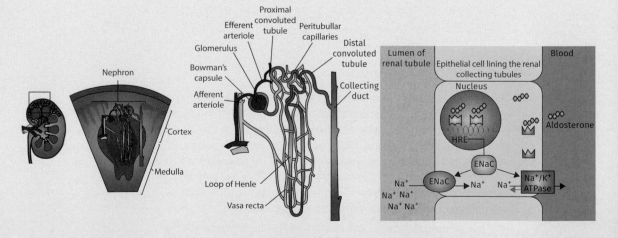

In this way aldosterone increases the ability of the kidney to conserve sodium when the body needs it. When the blood levels of sodium return to the physiological set level, a process of negative feedback stops aldosterone production and as a result no further sodium channels are produced. In this way a physiological overshoot is avoided, where too much sodium retention occurs.

It is worth noting that mineralocorticoid receptors are found in many different tissues of the body: the heart, large intestine, salivary glands, and the brain. The actions of aldosterone in these tissues are not quite as simple as its actions in the kidney.

❓ Pause for thought

Many advances in science have been the result of what is called 'serendipity', which partly means 'we got lucky'. The discovery of aldosterone is a great example of this because it came about as a result of a mistake. A husband and wife team, Jim and Sylvia Tait, at the Middlesex Hospital in London, were looking for the hormone that was known to control sodium reabsorption: aldosterone. They used a method of separating steroids by paper chromatography, where steroids would move along a paper strip at different speeds, depending on how hydrophobic they are. They had tried all other ways of separating steroids but had still not found aldosterone. Then, after one big experiment they forgot to stop the paper chromatography and got back to work the next morning to find it had run overnight. Almost certainly, anything of interest would have moved right up to the end of the paper strip: a complete disaster. But they analysed the paper strip anyway. Lo and behold, aldosterone was far less hydrophobic than any other steroid and this was the only way they could have found it. Would you have thrown away the paper strip and started again or would you have thought, 'We might find something interesting here?'

What do we mean when we say that an experiment has 'gone wrong'?

Switching off the hormonal signal

We have seen how hormone production is regulated by negative feedback. A physiological imbalance, such as low plasma sodium, results in an increased production of aldosterone, which acts on the kidney to stimulate sodium reabsorption. When the plasma level of sodium is restored, aldosterone production is no longer stimulated and so levels of this hormone fall. Once the hormone has triggered a physiological response in its target cell, how is that response terminated? All target cells have mechanisms for returning to the basal state, before their functions were changed by the hormone.

Peptide hormones, which bind to cell surface receptors often only stay bound for a short while and then detach. There is an equilibrium, or balance, between the hormone in blood and the number of receptors which are occupied at any one time. When the hormone detaches, the receptor returns to its resting state and the processes it has activated simply stop. Second messenger production falls back to basal levels and kinases become dephosphorylated. Sometimes peptide hormones, bound to their receptor, become internalized within the cell. When this happens, although the hormone remains attached, the receptor is no longer able to signal to the cell. The hormone–receptor complex is processed in a lysosome and the receptor can be recycled and returned to the cell surface, or degraded within the cell.

There are similar mechanisms in place for steroid hormone receptors. These use **chaperone proteins**, which are a group of proteins associated with the receptor, responsible for shuttling the receptor around the cell, binding to the DNA, engaging other transcription factors, and ultimately recycling the receptor.

These processes are not yet fully understood. However, we do know that when there are very high blood levels of a hormone for a prolonged time, there is a process of **receptor down-regulation**. This causes a decrease in the number of available receptors for a hormone and so reduces the biological response to the hormone. It can result in a form of hormone resistance, which we shall look at in Chapter 2.

Scientific approach 1.2
Modern methods in endocrinology

Berthold removed testosterone from his chickens by surgically removing their testes. Modern molecular methods mean that we can '**knockout' genes** involved in hormone production. For peptide hormones this is relatively simple. Replacing the gene for a peptide hormone with a mutated gene, or even deleting the gene completely, results in the production of an inactive peptide, or no peptide production at all. We can do the same thing with hormone receptors; as these are all proteins, the gene which codes for the receptor can be altered or deleted. These molecular methods were first introduced in the second half of the twentieth century but have been refined considerably since then. It is now possible to produce genetically engineered laboratory animals which have a gene 'knocked down' at a specific stage in development or in one specific cell type, leaving the other stages of life or tissue types unaffected. These engineered animals are very expensive to produce and the techniques are usually only used on mice, which are easy to breed and a very good model for studying.

In other animals, particularly non-mammalian species, different approaches are needed. If it isn't possible to breed genetically engineered animals with a disrupted gene, an alternative approach is to disrupt the process of translation or transcription. When a gene is transcribed a single strand of mRNA is produced, called 'sense RNA'. It is possible to construct a complementary strand of RNA that hybridizes with the sense RNA and stops it from being translated into protein. This complementary strand is called '**antisense RNA**'.

If the antisense RNA is injected into a specific brain region, for example, it can stop the product of a gene from being translated in that region for a certain amount of time; enough time to study the effects of blocking the gene. This is quite a precise method of effectively knocking out a gene.

The gene that is knocked out can be a gene coding for a peptide hormone or for a receptor. Knocking out receptors is a common method in endocrinology. This can be done using antisense technology as previously described, or by using a chemical that stops the hormone from binding to a receptor. Usually these chemicals work by occupying the hormone binding site on the receptor protein. A chemical that works in this way is called a **receptor antagonist**. Antagonists to steroid hormone receptors are often used to treat disease. We shall see some examples of this in Chapter 2.

❓ Pause for thought

Why does gene knock-out technology only work for peptide hormones and receptors? What methods might be used to look at the effects of removing steroid or thyroid hormones?

≋ Chapter summary

- Hormones are a range of different compounds, including peptides and steroids.
- Oversimplification of information about hormones can lead to some misleading ideas. In fact, not all hormones are slow acting, the endocrine and nervous systems are closely connected, and hormones are released by a range of different tissues in the body, not just the 'classical' endocrine organs.
- Steroid hormones are all made from cholesterol by a process called steroidogenesis. Peptide hormones are made just like any other protein.
- The effects of hormones are all as a result of the hormone binding to its receptor. These receptors are found in target tissues and their cellular location depends on the chemical type of hormone. Hormones all affect a wide range of target tissues, not just the 'classical' targets in the textbooks.
- Negative feedback control is the most effective means of regulating blood levels of hormones and their effects.
- The study of hormones has changed as more sophisticated methods have become available. There is still a lot that we don't fully understand about the endocrine system.

Further reading

Hinson, Raven, and Chew: *The Endocrine System* (Churchill Livingstone)
This is a textbook mainly aimed at medical students, which goes into more detail about the endocrine system.

https://www.hormone.org/hormones-and-health/hormones
The American Endocrine Society's online resource centre, listing all the major hormones with information about each one. This website relates to human health and hormones.

Discussion questions

1.1 As you will recognize, you have previously been given an extremely simplified version of the endocrine system. Do you think that we have given you everything there is to know about steroid hormone action, for example?

1.2 How can you go about finding out more detailed information about a topic that appears in this book? Which is the most reliable source of information: a textbook, a website, or an article published in a scientific journal?

2 HORMONES AND DISEASE: WHAT HAPPENS WHEN THINGS GO WRONG?

In Chapter 1, we saw how hormones are biological molecules with a wide range of actions in many different parts of the body. We also gained an insight into the lengths to which the body goes to ensure that hormone levels are 'just right' for the conditions at the time. So what happens when these very tight control systems go wrong? The answer is simple: when we produce either too much or too little of any hormone, then we usually become unwell. In this chapter we will look at the different types of endocrine disease, focussing on human health and disease. There are endocrine diseases in animals but these are not very different to the diseases we see in people.

All of the diseases associated with hormones result from either too much hormone activity or too little. The correct term for too much hormone activity is '**hyper-**' and for too little it is '**hypo-**'. So too much thyroid hormone activity is called hyperthyroidism and too little is called hypothyroidism (see Figure 2.1).

It is helpful to think about what we mean by hormone 'activity'. Too much hormone activity is usually the result of a disorder in the gland that produces the hormone, often due to uncontrolled growth of the gland. This means that the gland produces too much hormone. Too little hormone activity can be a fault in hormone production or it can be a problem with hormone action, where the hormone receptor doesn't work properly. We will look at examples of each of these types of endocrine disease.

Figure 2.1 The swollen neck of the woman in this painting from the fifteenth century shows a goitre—a sign of thyroid problems going back around 550 years!

Andrea Mantegna, *Madonna and Sleeping Child*, 1465–70. Photo: De Agostini Editore/ AgeFotostock.

2.1   Measuring hormones

First of all we need to understand how hormones are measured, because measuring levels of a hormone is a key step in diagnosing an endocrine disease. Hormones are extremely potent molecules: it only takes a tiny amount to have a very significant effect. Their concentration in the blood is usually very low indeed. We're familiar with the idea that blood glucose concentrations are about 5 mmol/l and electrolytes also circulate in millimolar concentrations, typically 140 mmol/l for sodium, for example. Hormones are found in concentrations more than a million times lower than this, at nanomolar and picomolar concentrations: see Table 2.1.

The very low concentration of hormones in the blood makes it particularly difficult to measure their levels accurately. But it is really important to be able to measure hormone concentrations accurately, because the difference between the appropriate level of a hormone and either an insufficiency or an excess is also very small. With hormones small changes can have big effects!

Individual hormones were identified and purified in the early part of the twentieth century. At this time, with the technology available, it was extremely difficult to measure hormone levels. The diagnosis of endocrine disease was mostly carried out, as it had been before the discovery of individual hormones, by the doctor observing physical changes in the patient's appearance and taking a careful account of the symptoms that the patient had noticed. Many endocrine diseases start with quite subtle changes and progress slowly so that, by the time a person goes to their doctor, the illness is often quite advanced. By the middle of the twentieth century doctors

Table 2.1 Concentrations of different substances in the blood

Substance	Type of substance	Concentration in the blood using conventional abbreviations	Log mol/l and equivalent SI unit (per litre) in full
Sodium	Electrolyte	140 mmol/l	10^{-1} 100 millimoles
Bicarbonate	Electrolyte	21–26 mmol/l	10^{-2} 10 millimoles
Glucose	Sugar	3–5 mmol/l	10^{-3} 1 millimole
Uric acid	Metabolite	150–500 µmol/l	10^{-4} 100 micromoles
Iron	Trace mineral	10–30 µmol/l	10^{-5} 10 micromoles
Vitamin A	Vitamin	0.5–2 µmol/l	10^{-6} 1 micromole
Cortisol (morning)	Hormone	200–650 nmol/l	10^{-7} 100 nanomoles
Testosterone (men)	Hormone	10–35 nmol/l	10^{-8} 10 nanomoles
Tri-iodothyronine (T3)	Hormone	1–3.5 nmol/l	10^{-9} 1 nanomole
Adrenaline (resting)	Hormone	170–500 pmol/l	10^{-10} 100 picomoles
Free (unbound) thyroxine	Hormone	10–30 pmol/l	10^{-11} 10 picomoles
Oxytocin (basal)	Hormone	1–4 pmol/l	10^{-12} 1 picomole

realized that if they were able to measure hormones, then they could detect too much or too little much sooner, and begin treatment before too much damage was done.

As we can see from looking at the history of pregnancy testing in Scientific approach 2.1, the introduction of immunoassays in the 1970s provided an excellent method for measuring hormones. 'Immuno' means using antibodies to 'recognize' and bind to the hormone and '–assay' means a method for measuring something. Immunoassays have the essential four characteristics of an analytical method: they are accurate, precise, sensitive, and specific.

- **Accuracy**: an accurate method means that the assay measures what it is supposed to measure within the range it is designed to work in.

- **Precision**: a precise assay is one that produces the same result, no matter how many times you repeat the measurement. The results of a precise assay are highly repeatable. In practice no method is 100% precise because there is always an element of human error and variation in measurement.

- **Sensitivity**: a sensitive assay is able to measure changes in hormone concentration within the range of hormone concentrations that are important for diagnosis. If an assay is not sensitive, then significant changes in hormone concentration may be missed (giving what is called a 'false-negative' result).

- **Specificity**: a specific assay means that there is nothing that is likely to interfere with the assay and change the result. In the early days of immunoassay some common drugs would affect the binding of the hormone to the antibody and so give a false-positive result. For hormones, specificity is a really important concept; we want assays to measure only the biologically active form of the hormone and not any chemically related substance, such as a metabolite that has no biological activity.

Scientific approach 2.1
History of pregnancy testing

The history of pregnancy testing during the twentieth century is a good example of how methods for measuring hormones changed with the availability of increasingly sophisticated technology. The modern pregnancy test is one of the simplest hormone measurements we carry out. It is based on the question: is the hormone **human chorionic gonadotropin (hCG)** present in the urine or not? hCG is only produced during pregnancy and it starts to appear in urine very soon after conception, so it is the ideal basis for a pregnancy test. There is no need for an accurate measurement of how much of the hormone is present—it is just a matter of whether it is there or not.

Nowadays it is easy to take for granted the availability of accurate pregnancy testing kits at an affordable price from any local pharmacy. A woman can tell if she is pregnant almost as soon as she misses a period. Throughout history there have been many attempts to devise a test which reliably confirms pregnancy. The earliest urine-based pregnancy test was devised long before people had heard of hCG, or indeed hormones. In ancient Egypt, about 3,500 years ago, there was a pregnancy test based on wheat germination. A woman's urine was sprinkled on wheat and barley grains. If the wheat germinated she was pregnant. Although it seems to be really unlikely, there is some scientific evidence to support this test. Scientists have found that 70% of the time it is accurate and suggest that it might be the high levels of oestrogens in the urine that caused the grains to sprout. A 70% accuracy rate isn't bad for 3,500 years ago. In medieval times there were 'piss prophets' who claimed to be able to detect a pregnancy by looking at, or smelling, urine. Some of these quack doctors did some very odd things, like burning cloth that had been dipped in urine and observing how it burned. Others mixed urine with wine and observed the characteristics of the mixture. You will not be surprised to learn that these 'tests' had almost no scientific basis.

In the early twentieth century, when the concept of hormones was being explored properly for the first time, a pair of German scientists were studying pregnancy. They discovered the presence of a substance, hCG, in the urine of pregnant women that caused development of the ovaries of immature rats. There was no method to directly measure hCG so they devised a pregnancy test based on injecting immature rats with urine, then killing the rats some time later and observing their ovaries to see if they had developed. The scientists were called Selmar Aschheim and Bernard Zondek and so their test was called the A–Z test. This pregnancy test was devised in 1927 and continued to be used for many years. It is an example of a **bioassay**: a method of detecting a hormone based on its biological effect.

These tests were expensive to carry out, slow to give results (several days in general), and required the killing of a large number of animals. A later version of a bioassay for pregnancy testing used a female toad: Xenopus laevis (see Figure A). The big advantage here was that toads respond to hCG by laying large numbers of eggs. There was no need to kill the toad to obtain the test result and so they could be re-used many times. There were still problems of both sensitivity (it takes quite a lot of hCG to make a toad ovulate) and specificity (the presence of another hormone, LH, in urine could have the same effect).

Figure A A woman would give a urine sample at the doctors, it was sent away to labs to be applied to toads like this *Xenopus* toad, and if they laid eggs, the doctor was informed and then gave her the news that she was pregnant. It seems incredible that this went on until the early 1970s!

Brandy McKnight/ Shutterstock.com

In the 1970s, tests for measuring hormones were devised, based on their immunological properties rather than their biological actions. These are called **immunoassays**. Immunoassays use antibodies which recognize the hormone and bind to it. They have the tremendous advantage that antibodies have a very high affinity for their targets and will bind to them even when a low concentration of the target is surrounded by hundreds of other chemicals. In order to be able to count the number of bound targets you need a marker, which is usually another antibody, this time with a measurable tag attached, which binds specifically to the target antibody. **Monoclonal antibodies** make the assays both more sensitive (able to measure lower concentrations of the hormone) and more specific (less likely to give false test results due to the interference of other substances). For an explanation of how immunoassays work, see Figure B.

The modern pregnancy test uses a technique called a sandwich **enzyme-linked immunosorbent assay (ELISA)** in a lateral flow system. This technique uses microfiltration methods which were only developed in the late twentieth century, combined with an immunoassay. The sandwich consists of the hormone hCG, with two antibodies attached to it. One antibody is used to detect the hormone and the second has a coloured tag which allows you to see the hormone. In a pregnancy test the hormone shows up as a coloured line on the test strip.

Figure B The principle of the modern pregnancy test. Urine (containing hCG) is applied to the sample pad. Capillary action draws it along the test strip towards the wicking pad. The test line shows hCG 'sandwiched' between two antibodies.

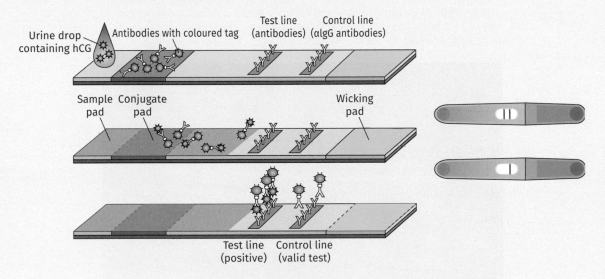

Ket4up/Shutterstock.com

Current home-use pregnancy test kits can reliably detect a pregnancy as early as around eight days after conception. As the ovum starts to produce hCG at between six and twelve days after fertilization, there is a high chance of false negatives (the assay being negative when the woman really is pregnant) if the test is carried out too early. Levels of hCG increase through the first months of pregnancy, so a later test is much more reliable.

❓ Pause for thought

Why do you think that LH could have the same effect as hCG on ovulation in Xenopus toads?

Patterns of hormone release and dynamic testing

As it became possible to accurately and reproducibly measure hormone levels in blood, scientists discovered that hormone concentrations could vary with the time of day. Some hormones vary in a consistent way, with high levels at one time of day and low levels at another, but always at the same times every day. This variation is called a circadian or diurnal rhythm. Not all hormones have a diurnal rhythm. It doesn't matter what time of day you do a pregnancy test, for example, because blood levels of hCG don't vary significantly during the day and, in any case, the pregnancy test simply determines whether or not hCG is present. Thyroid hormone has such a long half-life in blood that levels don't vary at all during the day. By contrast, the adrenal corticosteroid, cortisol, has a very clear diurnal pattern of secretion. Cortisol levels are high in the early morning, at around the time you wake up, and at their lowest while you are asleep.

Other hormones, such as insulin, are not released in a fixed daily rhythm, but in response to a change in the body. An increase in the concentration of glucose in blood causes an increase in insulin levels at any time of day or night. The pattern of growth hormone secretion is somewhere between insulin and cortisol; it has a clear diurnal rhythm with the highest levels seen during the early part of sleep but, like insulin, it is also affected by blood glucose levels. In this case, high blood glucose suppresses growth hormone secretion and so causes blood levels to decrease. On top of this, there is a randomness to growth hormone secretion, with peaks and troughs of hormone levels during the day (see Figure 2.2).

For the hormones like cortisol and growth hormone, finding the right time to test blood levels is difficult. It is important to ask the question: why are blood levels being measured? It makes a big difference whether the doctor is

Figure 2.2 The blood levels of cortisol and growth hormone have a daily pattern, called a diurnal rhythm. On top of this rhythm, there are also what appear to be fairly random peaks and troughs of hormone production during the day. This is why dynamic testing is needed.

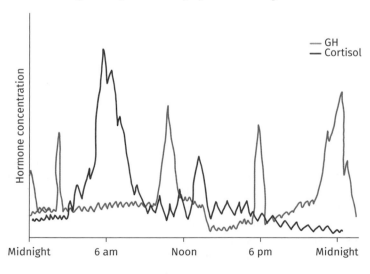

concerned about a possible lack of hormone or about a possible excess of hormone. As both time of day and eating can affect hormone levels, doctors have developed clear and specific protocols for testing each hormone. These protocols are also designed with the question in mind: do I think that levels of the hormone might be higher than expected or lower? If a doctor thinks that a patient might have a hormone deficiency, they will want to use a protocol that ensures the hormone levels are stimulated to be as high as possible. If they think that there might be a hormone excess, then they will try to lower hormone production. This might sound counter-intuitive, but it works.

Let us look at a specific example: growth hormone. Growth hormone is a large polypeptide hormone produced by the pituitary gland, which is controlled by a part of the brain called the hypothalamus. Growth hormone controls growth of bones and soft tissues like skin and muscle. As children it is particularly important that we produce the right amount of growth hormone because this drives the growth and development of healthy bones and all the other tissues in the body. If we have too little growth hormone, then we are not able to reach our full potential height. If we have too much growth hormone then we grow too tall. This condition is called giantism. You might think that getting tall wouldn't be a problem, but with excess growth hormone, production in childhood people can grow to over seven feet (2.1 m) tall. This can cause musculo-skeletal problems, but a much bigger problem is the effect of the hormone on the soft tissues of the body. Muscle and skin grows, but so do the heart and other internal organs, which can stop them from working properly. This is why it is important for health workers to monitor the growth of children and to investigate growth hormone levels if a child is growing either much too slowly or much too quickly.

If you look at Figure 2.2 you can see the problem with taking a single measurement of growth hormone. A blood sample taken at 1 pm would show an extremely low concentration of the hormone, but just an hour later the blood sample would show very high levels. In order to get around this problem there are test protocols which exploit the negative feedback regulation of growth hormone secretion.

If a doctor thinks that somebody might have excess production of a hormone, they use a test which aims to suppress hormone release. Growth hormone release is inhibited by high concentrations of blood glucose in healthy people. If there is a fault with the pituitary gland and it is producing excess hormone, the negative feedback effect of glucose doesn't work. A patient is asked not to eat or drink anything for a few hours before a test, then in the clinic they are given a glucose drink. One hour later they give a blood sample, which is tested for growth hormone. If the person doesn't have a problem with excess growth hormone production the glucose will suppress the hormone and there won't be any measurable growth hormone in their blood. If there is a measurable level of growth hormone, this suggests a problem and they go to the next stage of diagnostic testing.

If a doctor sees a child who is not growing as they should and is otherwise well, they might suspect low growth hormone levels and use a test designed to increase growth hormone release. Growth hormone release is increased by amino acids, particularly arginine. In this case the patient is asked not to eat or drink anything for a few hours before a test, then in the clinic they are given a drink containing the amino acid arginine (but *not* containing any glucose!). A blood sample is taken and growth hormone measured. If the person responds normally, the levels of growth hormone in their blood will be high after this test. If the growth hormone levels are still very low, or it cannot be detected at all, then doctors know there is a problem. In the past the growth hormone stimulation test used to involve giving insulin to lower blood glucose, in a reverse of the suppression test previously described. This was rather dangerous, as a very low blood glucose was needed to increase growth hormone and there was a risk of death. It is much safer to give an amino acid.

2.2   Autoimmune disease and hormones

The immune system works by detecting foreign cells or proteins in the body and producing big proteins, called antibodies, which bind to the foreign cell or protein and mark it for destruction by the cells of the immune system. Sometimes the immune system develops a fault and mistakenly identifies one of our own cells as foreign. When this happens, the immune system starts to produce antibodies directed against that particular cell type. These are called **autoantibodies** and their production leads to a condition which is called an **autoimmune disease**. The interaction of these antibodies with the cell causes a range of different problems in different cell types. Some well-known examples of autoimmune disease are coeliac disease and rheumatoid arthritis. Several endocrine diseases are also the result of an autoimmune response directed towards cells that produce hormones. These diseases include type 1 diabetes, thyroid disease (both over- and under-activity), and adrenal insufficiency. It's not clear why the immune system starts to do this, and particularly whether there is a trigger that initiates the problem. There is ongoing research to try to identify the triggers for these immune reactions because the disorders that result are both serious and long lasting.

Type 1 diabetes

Type 1 diabetes is an autoimmune disease. The immune system produces antibodies that bind to cell-surface proteins on beta cells in the endocrine pancreas. The bound antibodies effectively mark out the beta cells for destruction by the immune system, exactly as if they were an infective agent like a bacterium. As these cells are the source of insulin in the body, the destruction of beta cells means that no insulin can be produced. The result of this complete insulin deficiency is an endocrine disease called type 1 **diabetes mellitus**. It is called type 1 diabetes mellitus to distinguish it from both **type 2 diabetes mellitus** and from **renal diabetes insipidus**, which are not autoimmune diseases but diseases of hormone resistance.

Glucose is the main fuel for the cells of the body, used in cellular respiration to make ATP, which in turn is used to drive most of the chemical reactions which are vital for life. Without insulin the body cannot use glucose properly. When we have eaten a meal containing carbohydrates, there is an increase in blood glucose concentration. This usually causes an increase in circulating insulin, which works to enable cells, such as muscle cells, to take up this glucose from the blood to use in metabolism. Without insulin the cells can't take up glucose very well and the blood concentration of glucose remains high. Although the blood glucose is high, the cells in the body are not getting as much glucose as they need, because of the lack of insulin. So the body reacts as though blood sugar is low and hormones such as glucagon are released to try to increase the levels of glucose in the blood. As you can imagine, this only makes the blood glucose levels rise even higher. People who don't make insulin also get very tired because their cells aren't getting enough glucose to work at their best.

High blood glucose is a problem for several reasons:

- It damages blood vessels and nerves, especially the blood vessels and nerves in the hands and feet and in the eyes.

- It causes a problem with infections: there is nothing that bacteria like better than really sugary blood, so cuts and grazes are more likely to get infected and the infection is more likely to be serious.

- It causes kidney problems in two ways. The kidney has small blood vessels which are directly damaged by high blood glucose, but the kidney is also exposed to even higher glucose levels than the rest of the body. When the plasma is filtered by the kidney a lot of small molecules, including glucose, pass into the renal tubule in the filtrate. As this filtrate moves through the renal tubules this glucose is usually reabsorbed back into the blood and there is no glucose at all in the urine when it reaches the bladder. However, there is a maximum glucose level that the kidney can handle. This is called the renal threshold. When plasma glucose concentrations rise above the renal threshold, the kidney can't reabsorb all the glucose and so the urine contains glucose. Glucose has a high osmotic potential and so needs to be diluted with water. This means that the kidneys produce a lot of dilute, but sugary, urine. This in turn means you get very thirsty and need to drink a lot more fluids. Very often the first signs of type I diabetes are tiredness, needing to pee very frequently and feeling much more thirsty than usual.

Unlike most other endocrine disorders, the diagnosis of type I diabetes does not involve measuring the hormone involved. Instead, doctors who think that a patient might have diabetes ask the person to produce a urine sample. They test the urine for glucose using a simple test strip, and if glucose is present in the urine move onto the next step in diagnosis. This is to ask the person to fast (not eat anything) overnight and then take a blood sample in the morning. In a healthy person the blood glucose should be within a narrow range (3–5.5 mmol/l). If it is higher than this, then the person has a form of diabetes. The blood can also be tested for the presence of autoantibodies against parts of the pancreatic beta cells. The most common autoantibodies bind to cell-surface markers on the beta cells, or to a zinc transporter protein specific to beta cells, and occasionally to insulin itself. The presence of one or more of these antibodies confirms type 1 diabetes.

At the moment the only treatment option for type 1 diabetes is to give insulin (see Chapter 6 for more information about how the insulin is made). Scientists are working to develop a stem cell therapy. They are trying to engineer stem cells to develop into fully functional beta cells. This would mean that the cells would have to be able to detect blood glucose levels and produce the right amount of insulin in response to changes in blood glucose, just like healthy beta cells. These engineered beta cells would also need to have modified cell-surface proteins, so that they were not recognized as foreign by the immune system. This is the real challenge of research into type 1 diabetes. Another approach is to use new gene-editing technologies such as CRISPR/Cas9 to edit the genomes of beta cells in affected individuals to prevent the immune response and enable the cells to survive, reproduce, and produce insulin again. The hope is that in the future, instead of treating type 1 diabetes, we will be able to cure it.

Thyroid disease and autoantibodies

One of the most common endocrine targets for autoimmune disease is the thyroid gland. Studies have shown that around 20% of adult women have anti-thyroid antibodies, although not all of these antibodies cause disrupted thyroid function. It is a curious thing that autoimmune disease of the thyroid can cause both hyperthyroidism and hypothyroidism: overactivity and underactivity of the gland. Both of these conditions cause health problems. We saw the effects of an underactive thyroid in Chapter 1 and we will discover the effects of excess thyroid hormone in Chapter 6. The outcome depends on which thyroid-specific protein the antibodies recognize.

To understand autoimmune thyroid disease, we need to look at how thyroid hormones are made. The thyroid gland is made up of clusters of cells surrounding a pool of colloidal protein called **thyroglobulin**. The thyroid cells synthesize the protein, add iodine residues to it, and secrete it into the colloidal pool. The thyroid hormones are made by **endocytosis** of the protein and processing within the thyroid cells. The hormone is then released into blood. You can see the structure of the gland in Figure 2.3.

There are three different proteins which are the target of autoantibodies in most cases of autoimmune thyroid disease: the TSH receptor, thyroglobulin, and an enzyme, thyroid peroxidase. A person with autoimmune thyroid disease will usually make antibodies to just one of these proteins, but these are the three most common targets. These proteins are all found specifically in the thyroid and nowhere else in the body, so the antibodies produced only affect the thyroid gland. Occasionally there are antibodies to the other thyroid-specific proteins on the cell surface, including the sodium–iodine transport protein that is responsible for taking up iodine into thyroid cells.

The hyperthyroid version of autoimmune thyroid disease is called Graves' disease, named after the nineteenth-century doctor who first described the condition. This is the most common form of thyroid disease. In Graves' disease the autoantibodies bind to cell-surface TSH receptors on the thyroid hormone-producing cells and activate hormone production. The activated

Figure 2.3 (a) The thyroid gland is located in the throat, with a lobe on either side of the trachea (windpipe). It is often described as being like a bow-tie. (b) If you look at a section of thyroid gland under the microscope you can see that thyroid cells surround the colloidal protein.

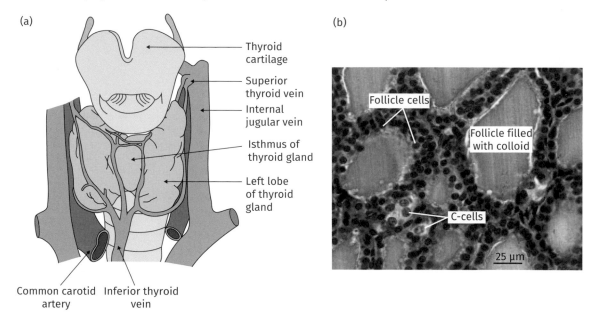

(a)
- Thyroid cartilage
- Superior thyroid vein
- Internal jugular vein
- Isthmus of thyroid gland
- Left lobe of thyroid gland

Common carotid artery   Inferior thyroid vein

(b)
- Follicle cells
- Follicle filled with colloid
- C-cells
- 25 μm

Courtesy Christopher D. Richards

receptors also make the thyroid gland grow larger. Sometimes the first sign that somebody notices is that they have a swelling in their neck. This swelling is called a goitre (pronounced goy-tuh)—look back to Figure 2.1 to see an example. Having a goitre does not mean that the thyroid is overactive; sometimes it means the opposite. If somebody doesn't get enough iodine in their diet to make the right amount of thyroid hormone, the thyroid gland can grow larger as part of the homeostatic response to thyroid hormone deficiency. It gets bigger and bigger in an attempt to make more hormone.

The hypothyroid form of autoimmune thyroid disease is called Hashimoto's thyroiditis. The antibodies cause inflammation of the thyroid gland, which swells and becomes visible as a goitre. As the disease progresses the gland can be destroyed, leaving a thyroid hormone deficiency. This is similar to type 1 diabetes mellitus; the autoantibodies bind to cell-surface proteins, marking out the cells for destruction by the immune system (see Figure 2.4).

Addison's disease and adrenal failure

The adrenal gland (see Figure 2.5) is a fascinating endocrine gland. It effectively consists of four glands all rolled into one and arranged in concentric circles. The middle of each adrenal gland is the **adrenal medulla**. This is the part that is involved in the 'fight-or-flight' response and produces adrenaline. Around this middle part are the three layers of the **adrenal cortex**, each producing a different set of steroid hormones. The most important

Figure 2.4 Thyroid homeostasis and the development of goitre—both under- and over-stimulation can lead to the gland getting bigger

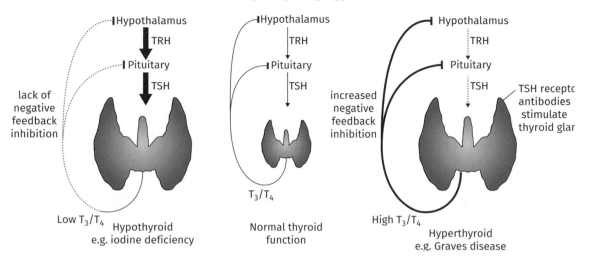

Figure 2.5 Structure of the adrenal gland

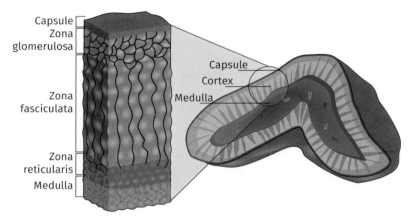

Sakurra/Shutterstock.com

adrenal steroids are called **glucocorticoids**, the most important glucocorticoid hormone being cortisol. Glucocorticoids were named because they were thought to be important in glucose metabolism. While this is true, it is only a small part of their function. The glucocorticoid receptor, which binds cortisol, is found in almost every tissue and cell type in the body. In the liver alone there are 1,300 genes that are regulated by cortisol. We now know that cortisol is important in the control of development, metabolism, and in the immune response. Without glucocorticoids, we die.

Autoimmune antibodies can destroy the adrenal cortex, which means that the body can't produce glucocorticoids. This isn't a sudden event, like having a heart attack. It happens slowly, like many endocrine disorders, and can be difficult to spot. As the antibodies attack the cells of the adrenal gland, cortisol production starts to decrease. This triggers the pituitary gland to produce more of the hormone that controls cortisol production, **adrenocorticotrophic hormone (ACTH)**. When ACTH levels increase, the cells of the adrenal cortex respond by getting larger and producing more cortisol. This allows the blood concentration of cortisol to remain quite stable, even though there are fewer and fewer adrenal cells making cortisol. The high levels of ACTH in the blood don't have any effect other than to increase the amount of cortisol, so at this stage, even though the system is stretched to breaking point, the disease may not have any noticeable symptoms. Sometimes, however, there is a very odd symptom indeed: some parts of the skin, especially the parts not usually exposed to the sun, such as on the palms of the hands, around the genitals, the inside of the mouth, or old scars, become darker in colour as a result of increased pigmentation. To understand the link between adrenal failure and skin pigmentation we need to look at the production of the hormone that controls cortisol release, ACTH.

The pituitary gland, at the base of the brain, produces ACTH in response to a signal from the hypothalamus. ACTH is a peptide hormone, made from a very large precursor peptide, called pro-opio-melano-cortin or POMC. This large peptide is processed to produce ACTH and a range of other peptides, including **melanocyte stimulating hormone, MSH**. This is the hormone that helps some animals to adapt their skin colour to blend in with darker backgrounds; it stimulates production of melanin, which causes the skin to darken. When ACTH production is very high, there can also be high levels of MSH, which increases melanin production in human skin, making the skin darker.

As the autoimmune destruction of the adrenal gland progresses and there are fewer and fewer cells remaining, cortisol levels in the blood begin to decrease. This only happens very late in the illness; you can lose 90% of the adrenal cells before getting any symptoms of adrenal insufficiency. Even at this stage the illness can be difficult to spot. People with adrenal insufficiency describe feeling generally unwell, with stomach pains or headache, a general feeling of tiredness, feverishness, and often nausea. If you think about how you feel when you come down with something as common as flu, you can see that these symptoms don't immediately suggest that someone might have a problem with their adrenal gland. Often, the first time anyone knows that they have insufficient cortisol is when they do get the flu. The body can just about cope with a lack of cortisol until there is a stress on the system, such as a viral infection, which can tip the system over into a much more obvious disease state. This is called a hypo-adrenal crisis or Addisonian crisis.

In a hypo-adrenal crisis the person feels very sick indeed and often vomits. Their blood pressure falls dramatically and they collapse. If they aren't treated they can become unconscious through dehydration and low blood pressure and will die. Although this is a rare condition, doctors know to look out for it and it is very quickly and easily treatable, once it is identified.

Case study 2.1
Thomas Addison 1793–1860

Figure A Thomas Addison—the doctor who first recognized the effects of a complete failure of the adrenal gland on his patients

Wellcome Collection

The first person to describe primary adrenal insufficiency (most commonly caused by autoimmune destruction of the adrenal gland) was an English physician called Thomas Addison (see Figure A). He was mainly interested in anaemia but in 1855 identified a new disease. He described a group of eleven seriously ill patients at Guy's Hospital in London. These patients died and he performed a post-mortem examination on them. He described what he found when he examined the patients before and after death:

> *The discoloration pervades the whole surface of the body, but is commonly most strongly manifested on the face, neck, superior extremities, penis, scrotom, and in the flexures of the axillae and around the navel . . . The leading and characteristic features of the morbid state to which I would direct your attention are, anaemia, general languor and debility, remarkable feebleness of the heart's action, irritability of the stomach, and a peculiar change of the colour in the skin, occurring in connection with a diseased condition of the suprarenal capsules (adrenal glands).*

The condition he described is called **Addison's disease,** now more commonly known as primary adrenal insufficiency. Around 80% of cases of primary adrenal insufficiency are a result of autoimmune disease.

 Pause for thought

Bearing in mind the principles of dynamic testing for insufficiency or excess that we discussed earlier in this chapter, how would you design a test for Addison's disease?

2.3   Problems with steroid hormone production: inborn errors of metabolism

Steroids are made from cholesterol by a series of reactions catalysed by enzymes. Enzymes are proteins and the genes which code for them can mutate. When this happens, the enzyme works less efficiently and the pathway of steroid production can be disturbed. We have just seen what happens when the adrenal gland fails to make the corticosteroid cortisol in Addison's disease. When there is a mutation in the gene for one of the enzymes in the pathway for cortisol production, this can cause a condition called congenital adrenal hyperplasia, or CAH. The condition is congenital because it is something that you can be born with, it doesn't develop during life. It affects the adrenal glands and is called hyperplasia because of the growth of the adrenal glands caused by increased cell division. If we look at the pathway of steroid production and the control of adrenal gland function we can understand how this disorder happens (see Figure 2.6).

As you can see from the figure, the enzyme defect means that the adrenal gland makes less cortisol than the body needs. The decreased negative feedback from reduced levels of cortisol causes additional ACTH release, with the effect that the adrenal gland is overstimulated by ACTH. The bottleneck in cortisol production causes the adrenal gland to produce a lot of androgens. CAH is a disorder that starts before a child is born. You can find out more about the effects of androgens on the unborn child in Chapter 3. Here we should just note that a child with CAH is usually quite unwell and doesn't feed and grow as it should, until it gets treatment. One clue that there is a problem can be that a baby is born with pubic hair because of the high levels of androgen. A girl born with CAH often has an enlarged clitoris, which can be mistaken for a small penis; sometimes the midwife won't be sure whether the baby is a boy or a girl. We will look much more closely at the effects of androgens on development in Chapter 3.

Figure 2.6 Pathway of steroid production in the adrenal cortex. Each step is catalysed by a different enzyme. If there is a defect in one of the enzymes of the cortisol and aldosterone pathway, this causes a deficiency in these hormones. It also means that the pathway backs up and there is overspill into the androgen pathway. A relative lack of cortisol means that the pituitary gland releases more ACTH to stimulate cortisol production. This makes the adrenal gland grow and produce even more androgens.

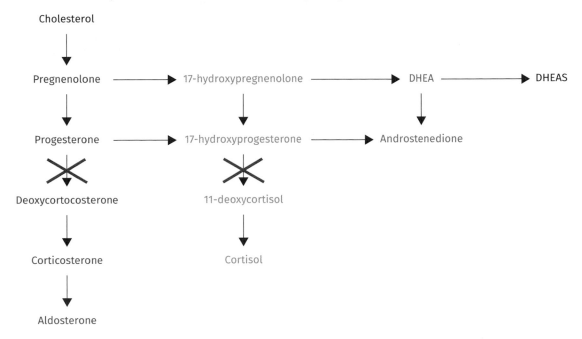

2.4　Hormones and cancer

To most people the word *tumour* means cancer. This isn't always the case. A tumour is a growth, caused by the uncontrolled proliferation of cells. Only if it spreads into other areas of the body do we describe it as cancer. In the endocrine system we usually find localized tumours that aren't malignant (invasive), so are not cancers. These tumours produce hormones in an unregulated way and are called **adenomas**. They are normally discovered during medical investigations because these adenomas often only show up when they cause symptoms of an excess of whichever hormone the cell type usually produces. Adenomas are treated by surgical removal of the tumour.

Sometimes a hormone-producing tissue can become cancerous, but this is relatively unusual. The most common endocrine cancer is found in the thyroid gland. Thyroid cancer is the fifth most common cancer in the USA and is much more common in women than men. What happens is that a group of thyroid cells become altered slightly so that they are no longer dependent on the usual control from TSH (thyroid stimulating hormone). They effectively go rogue and start growing out of control, producing large amounts of thyroid hormone. Unlike the autoimmune thyroid diseases that we have just looked at, thyroid cancer doesn't cause a goitre, but can cause all the symptoms of thyroid hormone excess.

Occasionally a cancer in a different sort of tissue can suddenly start producing a hormone. This is a very odd thing indeed and is called **ectopic hormone production**, meaning that the hormone comes from somewhere completely different to the cell type that usually makes it. An example of this is a type of lung cancer, called small cell carcinoma, which can produce several different types of hormone. This is very unusual indeed.

The most common relationship between hormones and cancer is that some types of cancer depend on hormones to make them grow. Prostate cancer is usually dependent on the presence of an androgen such as testosterone, while some types of breast cancer depend on oestrogens (called oestrogen-dependent breast cancers). These oestrogen-dependent breast cancers grow when oestrogen is present and so one effective form of treatment is to remove the oestrogen, so that the growth of the tumour slows or stops. In the nineteenth and early twentieth centuries doctors treated these cancers by surgically removing the source of oestrogens: ovaries in younger women and the adrenal glands in older women. This drastic action caused a number of serious problems, as you can imagine. In the past it was not possible to tell before surgery whether a breast cancer was oestrogen dependent, so a lot of unnecessary operations had to be carried out, just to be on the safe side. Nowadays a biopsy of a breast cancer can be quickly tested to see if it is oestrogen dependent, and a lot of the invasive surgery has been replaced by drug therapy.

We saw in Chapter 1 how it is possible to remove the action of a hormone by using a chemical to stop the hormone binding to its receptor. Oestrogen receptor antagonists can be used to treat oestrogen-dependent cancers, but the problem is that oestrogens are important in many different tissues of the body, particularly the brain and bone. What was needed was a type of oestrogen receptor antagonist that blocked oestrogen action in breast tissue without affecting bone and other tissues. Scientists discovered that the oestrogen receptors found in different tissues have quite different properties. This has led to the development of a set of drugs called selective oestrogen receptor modifiers (SERMS), which are used to treat hormone-dependent breast cancer. The best known of these drugs is tamoxifen. Tamoxifen is an oestrogen receptor antagonist in the breast, so it blocks oestrogen action in the breast. This makes it a good treatment for breast cancer and it is widely used to prevent any recurrence. In bone, however, tamoxifen activates oestrogen receptors, and so is acting as an oestrogen receptor agonist. This means that it is able to prevent damage to bone that may result from lack of oestrogen action. As you can see, the science of hormone receptors is rather more complicated than it first appears.

2.5   Hormone resistance syndromes

All hormones act by binding to and activating their own specific receptors. As we saw in Chapter 1, these receptors are large, complex proteins and hormones bind to them by interacting with a special binding region of the protein. The shape of this binding region is critically important and depends on the tertiary structure of the protein. Any mutation which alters the amino acid sequence of the protein is likely to affect the tertiary structure and impact on hormone binding.

One of the more unusual, but very interesting, conditions caused by hormone resistance is renal diabetes insipidus. Diabetes insipidus means *weak urine*. People who have this condition produce huge amounts of extremely dilute urine. They are constantly thirsty but never seem able to satisfy their thirst because of the constant urine production. Adults usually produce less than three litres of urine each day. People with diabetes insipidus produce up to fifteen litres every day; over five times the usual amount! The condition is caused by lack of vasopressin action. This can either be a result of a problem with the production of vasopressin, or, in the case of renal diabetes insipidus, by a problem with its action.

In the kidney, vasopressin has its effects by binding to a receptor called V2. When vasopressin binds to this receptor it causes activation of a water transport protein called aquaporin 2. It is the activation of this protein that allows water to move from the urine, through the kidney tubule, and into the blood.

In renal diabetes insipidus a defect in the vasopressin V2 receptor means that it does not respond properly to vasopressin. The hormone still binds to the receptor, but doesn't have the same effect on the target cell. This resistance to vasopressin is usually a condition that develops during life rather than a genetic disorder that somebody is born with. It is not possible to treat by giving more vasopressin and so is treated by controlling the amount of water and salt in the diet. Left untreated, the large amount of urine produced, even though it is very dilute, results in very low blood sodium concentrations, causing neurological problems and eventually death. It is a very unusual but quite remarkable condition, illustrating the tremendous job that the kidney normally does in retaining water.

Type 2 diabetes mellitus and the metabolic syndrome

Type 2 diabetes is a condition where the body produces insulin, but not enough to maintain a normal fasting blood glucose level. One of the main causes of type 2 diabetes is insulin resistance, where insulin is produced normally, but the cells of the body do not respond to it. As a result, the insulin doesn't have the same effect that it does in healthy people. Type 2 diabetes is very common indeed: recent data from the USA estimate that 25% of the American population has insulin resistance.

Insulin resistance means that there is a disconnection between the level of insulin in the blood and the biological effect on blood glucose levels. In other words, a significantly greater amount of insulin is required to keep blood glucose levels within the normal range in people with insulin resistance compared to those who do not have the condition. Insulin resistance is not a result of a mutation in the receptor or drug-related damage to the receptor. Like diabetes insipidus it is an acquired condition, but giving increased amounts of insulin or causing an increase in insulin production is an effective treatment. Insulin resistance is most often seen as part of a condition called the 'metabolic syndrome'. It seems that this condition is triggered by a combination of obesity (see Figure 2.7) and lack of exercise, which both affect the body's ability to take glucose out of the blood and

Figure 2.7 An increase of obesity has led to a parallel increase in the incidence of type 2 diabetes

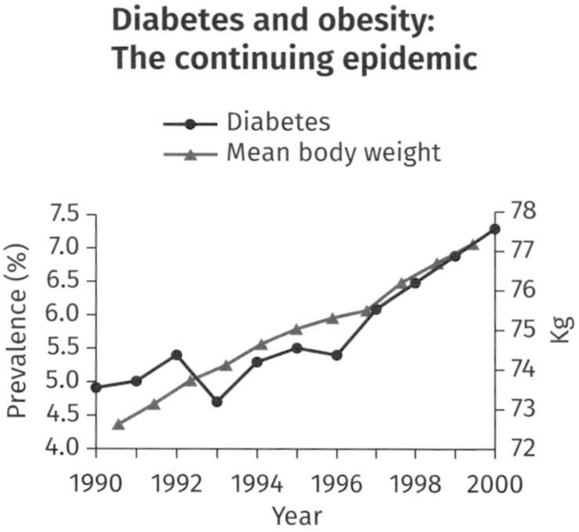

Left: Roberto Michel/Shutterstock.com

into cells. The metabolic syndrome is a group of metabolic and physical changes that includes insulin resistance, high blood lipid (cholesterol and triglyceride) levels, increased fat around the waist, and high blood pressure. People with metabolic syndrome have a greatly increased risk of heart disease, stroke, and early death.

The mechanism of the insulin resistance is not completely understood, although several theories have been proposed. It has been suggested that changes in membrane fluidity affect the way that the insulin receptor is located in the plasma membrane. This could affect its ability to signal to the cell. Insulin binding to its receptor causes an increase in functional glucose transporter proteins at the surface of the target cell. There are many different steps in between the insulin binding to the receptor and the increase in glucose transporters at the cell surface, so very many places where the process could be disrupted. Type 2 diabetes is what is called *multifactorial*, which means that many different factors contribute to the development of the disease. There is some genetic susceptibility; some groups of people are much more likely to develop this disorder than others. In Fiji and other South Pacific islands, for example, there is a much higher incidence of type 2 diabetes than in Western Europe.

In most of the developed world lifestyle is the main factor that determines insulin resistance and type 2 diabetes. When people become obese and take less exercise, their cells become more resistant to insulin. The pancreas compensates by releasing extra insulin to overcome the resistance. Eventually the pancreas can fail completely as the diabetes progresses, and people with type 2 diabetes can become dependent on injected insulin. Before the disease reaches this point, however, there is a lot of evidence to show that a majority of people can completely reverse the disease process by reducing their body weight and taking exercise. We don't understand the mechanism, but it is clear that exercise is a good way to boost the effectiveness of insulin.

It used to be possible to tell the difference between the two forms of diabetes by calling type 2 diabetes 'adult-onset', reflecting the fact that it was only in adulthood that this disorder developed. With the huge increase in the number of obese children and adolescents, this label no longer works. The increase in the number of obese people has been called an *epidemic*. It has been predicted that in countries such as the USA, where levels of obesity are particularly high, the average life expectancy will soon start to fall significantly. The good news is that losing weight and taking moderate exercise is effective in delaying or reversing this condition in more than half of people.

Very recently diabetes mellitus has been reclassified into five forms of the disorder, based on a genetic analysis of a large number of patients. This has allowed scientists to define five groups or clusters of patients with a diagnosis of diabetes mellitus. The first form, called cluster one, is *severe autoimmune diabetes* causing a lack of insulin production, which describes the previous classification of type 1 diabetes. The other four forms are sub-classifications of type 2 diabetes. Cluster 2 is called *severe insulin-deficient diabetes* with insulin deficiency not caused by autoimmune disease and is seen in people who are not obese. Cluster 3 describes a severe insulin-resistant state, mostly seen in people who are overweight. Cluster 4 is a mild insulin-resistant state, again seen in people with obesity, while cluster 5 describes a mild, age-related diabetes, associated with people who are older than those in cluster 4.

Laron syndrome

Growth hormone is produced by the pituitary gland and, as its name suggests, it is important in controlling growth, especially of the long bones of the arms and legs. Laron syndrome is a condition of growth hormone receptor insensitivity. Laron syndrome is an extremely rare condition and affects only about 350 people across the whole world. Like many genetic disorders, it is associated with a particular group of people; about one-third of all the people who have this condition live in a particular part of Ecuador, in South America. In Laron syndrome, a mutation in the receptor for growth hormone means that the hormone cannot bind to the receptor and so growth hormone cannot act. People with Laron syndrome have a complete lack of growth hormone action. At birth, there is usually no difference in the baby's length or weight, as growth hormone is not particularly important for controlling growth in the womb. However, growth hormone is the main hormone responsible for growth during childhood and also plays an important role during the adolescent growth spurt at puberty. People with Laron syndrome, where the cells cannot respond to growth hormone, have very short stature both during childhood and as adults. Typically men with this condition grow to no more than 4 ft 6 in (137 cm) and women to no more than 4 ft (120 cm) (see Figure 2.8).

Figure 2.8 Drs Jaime Guevara-Aguirre (top left) and Valter Longo (top right) with some of the people affected by Laron syndrome, who are helping with the research into how this genetic change delivers such health benefits along with short stature

Reproduced with permission from Peter Bowes

Short stature is not the only feature of Laron syndrome: people with this condition are much more likely than the rest of the population to be obese. On the other hand, they are far less likely to develop either type 2 diabetes or cancer, which seems very surprising as both these conditions are generally far more common in obese people. This protective effect of Laron syndrome is why it is a fascinating condition for scientists to study, even though it is so rare. Scientists are studying the genetics of Laron syndrome to try to work out the link between the growth hormone receptor defect and the protection against cancer and type 2 diabetes. This research on just 350 people could have an impact on billions of people in the future.

Androgen resistance

Androgen resistance or androgen insensitivity (they mean the same thing) is a condition that affects about one in ten thousand people who are genetically male (Karyotype 46XY). About half of the people affected have a complete androgen insensitivity, while the other half have a partial insensitivity. It is an inherited condition, the result of a mutation in the DNA which codes for the androgen receptor. We will look at the developmental consequences of androgen insensitivity in Chapter 3.

Chapter summary

- Illness can be caused by having too much or too little of a hormone. Often the illness progresses very slowly and its effects are only seen when it has been present for a long time.
- Hormones are present in only very low concentrations and often the hormone concentration changes in an unpredictable way, so designing tests to measure hormones is challenging.
- Endocrine disease can have many causes, including autoimmune disease, defects in the receptor, and enzyme defects.
- Some forms of cancer are dependent on hormones and can be treated by blocking either production or action of the hormone.
- The application of genetic analysis to endocrine disease allows a better understanding of the underlying mechanisms of disease. This allows the development of more appropriate forms of treatment.

Further reading

https://history.nih.gov/exhibits/thinblueline/timeline.html
The USA National Institute of Health (NIH) website, full of really good information. This is the link to their account of the history of pregnancy testing.

Hinson, Raven and Chew: *The Endocrine System* (Churchill Livingstone)
This is a textbook mainly aimed at medical students, which goes into more detail about the endocrine system.

http://www.diabetes.org/
The American Diabetes Society website, which has a lot of very good information about all forms of diabetes.

https://www.hormone.org/hormones-and-health/hormones
The American Endocrine Society's online resource centre, listing all the major hormones with information about each one. This website relates to human health and hormones.

Discussion questions

2.1 Given the trend for weight to be increasing in the developed world, and the health risks associated with this, how would you set up a health campaign to address the obesity epidemic?

2.2 Think about the four essential characteristics of an analytical method (accuracy, precision, sensitivity, and specificity) in relation to the ELISA pregnancy test. What effect does each of these characteristics have on the usefulness of this method? For each characteristic work out what impact it would have on the usefulness of the test if the characteristic were not properly met.

3 HORMONES AND DEVELOPMENT

In evolutionary terms, the only measure of success is whether an organism reproduces successfully and replicates itself. In human society, in many places in the world, and for many individuals, the most important thing in life is to have children. But young animals, whether they are caterpillars, tadpoles, kittens, or human babies cannot reproduce. They have to undergo major changes to reach an adult, sexually mature form that can produce offspring—and this is where hormones come in.

In Chapter 1 we looked at Berthold's experiments on the development of adult characteristics in cockerels. In this chapter we will look at sexual development from conception to adulthood in a range of species, including humans. We will then look at what happens when chemicals that interfere with the endocrine system get into the environment and the impact that this can have on development.

In this chapter we use the word 'development' to mean the usual pattern of change that takes place during an organism's lifecycle from conception through to the final adult, reproductively active form of the organism (see Figure 3.1). The first organism we will look at is ourselves.

Figure 3.1 These lambs have to undergo many developmental changes to become like their parents—adult sheep capable of producing a new life. These changes will be driven by hormones.

3.1   Human sexual development

We know that in humans our biological sex is determined by our sex chromosomes at conception. Having two X chromosomes usually produces a female baby with the external genitalia that you would expect: a clitoris, a vagina, and labia. Having one X and one Y chromosome produces a male baby with a penis, scrotum, and testes. That all seems to be very straightforward and yet, as we see so often in biology, it is rather more complicated than that and the process of sexual development can go wrong. As with many other areas of human biology, it is the rare cases when things do not follow the expected route that can tell us a lot about how sexual development works normally.

For example, it occasionally happens that a baby with one X and one Y chromosome is born looking like a baby girl, and a baby with two X chromosomes is born looking like a baby boy. This complete disconnection between genotype (genetics) and phenotype (appearance) is unusual and it is rather more common for a baby to be born with ambiguous genitalia: looking neither quite like a baby boy nor a baby girl. In all of these cases there are a number of reasons why the development of the baby did not follow the expected plan.

To understand the development process we need to look at what happens between conception and birth. The foetus starts out with what is called a *primitive* or *indifferent* gonad, which describes a tissue that has the potential to develop into either a testis or an **ovary**. The path of development is controlled by hormones, particularly the male sex hormones, the androgens. If the foetus doesn't produce androgens then the primitive gonad develops into ovaries. If it does produce androgens, then the primitive gonad develops into testes. The default is always for a female baby to develop. So clearly we need to understand what makes a foetus produce androgens and override the default. To do this we need to look at the sex chromosomes.

The X chromosome is dramatically larger than the Y chromosome, and contains about 1,000 genes, compared to fewer than one hundred on the Y chromosome. However, there is an important gene on the Y chromosome called *SRY*, or sex-determining region Y. It is this gene that switches on the cascade of gene activation that leads the primitive gonad to develop into a testis. This happens early in the development of the foetus, at around six to seven weeks after conception. As the testes develop they start to produce androgens which inhibit the development of female genitalia and instead stimulate the development of a penis and scrotum. Androgens act on a structure called the tubercule to develop it into a penis. If you don't have androgens then the tubercule develops into a clitoris. These androgens also cause changes in the brain so that the baby, as he develops, feels like a boy.

So how can it happen that a baby with XY sex chromosomes can develop female genitalia and a baby with XX sex chromosomes can develop the appearance of male genitalia?

Mutations in the *SRY* gene that result in a failure to activate testicular development is one cause of a karyotypic male (46XY) with a female phenotype. This is called Swyer syndrome and is a rare condition, seen in only one in 80,000 births. At birth the baby is identified as female and as she develops she feels female. At the time of puberty it becomes clear that something isn't right because she doesn't go through all the usual development stages of puberty that her contemporaries will experience. She does develop pubic hair and her breasts may start to develop a little, but she never starts to menstruate. This is because her ovaries have not developed properly and have become nothing more than small clusters of fibrous tissue, so she will be infertile. In other words, you need a fully functional Y chromosome to become male, but complete development of the ovaries needs the presence of two X chromosomes.

The bigger picture 3.1

Karyotype and phenotype in human sex development

Figure A shows the karyotypes of a human female 46XX and male 46XY. This is the typical karyotype of males and females which gives the phenotypes of a man and woman, respectively.

However, the sex chromosomes are not always neatly paired. About one in every 200 female babies is born with only one X chromosome (karyo-

Figure A These young people are both karyotypically and phenotypically male and female, respectively—but our understanding of how we develop comes largely from individuals where the karyotype and the phenotype do not match

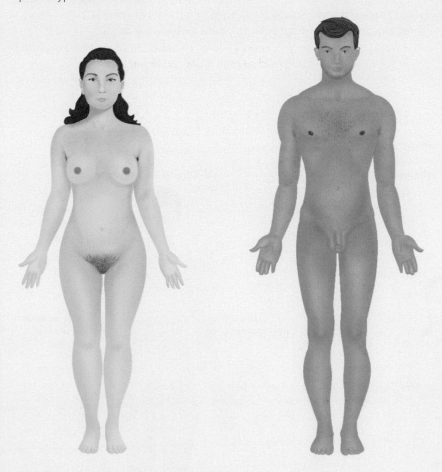

Jacopin / bsip / science photo library

type 45X0). This karyotype results in the phenotype of Turner syndrome (see Figure A). Around one in every 500 baby boys is born with an extra X chromosome (karyotype 47XXY). This additional chromosome results in the phenotype of Klinefelter syndrome (see Figure B). Women born with Turner syndrome are not able to have children, although men with Klinefelter syndrome may have reduced fertility, rather than being totally infertile. Around one in every 1,000 baby boys is born with an extra Y chromosome (karyotype 47XYY), which does not usually affect fertility but does result in a lower IQ.

Most of our understanding of the impact of the Y chromosome on development in humans has come from studies on people who have a phenotype that does not correspond to their karyotype.

❓ Pause for thought

Why is there not a syndrome with the karyotype 45Y0?
    We have seen what happens when somebody is born with three sex chromosomes. Do some research and find out what happens if they have more than this.

Figure B It is relatively common for somebody to be born with just one sex chromosome (Turner syndrome) or an additional X chromosome (Klinefelter syndrome). Both these genetic disorders have an impact on phenotype and fertility.

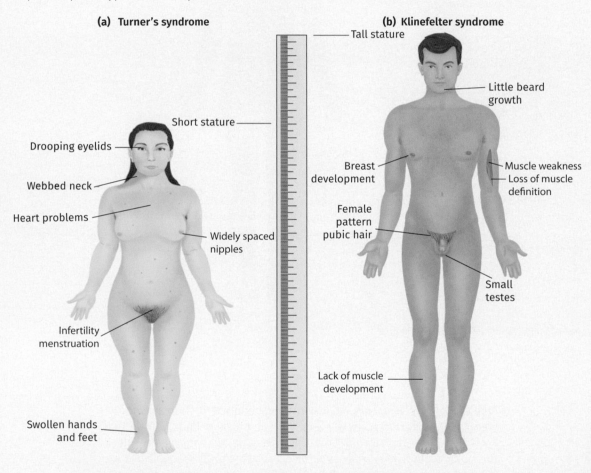

(a) Turner's syndrome

Short stature

Drooping eyelids

Webbed neck

Heart problems

Widely spaced nipples

Infertility menstruation

Swollen hands and feet

(b) Klinefelter syndrome

Tall stature

Little beard growth

Breast development

Muscle weakness
Loss of muscle definition

Female pattern pubic hair

Small testes

Lack of muscle development

Androgen insensitivity describes the inability of tissues in the body to respond to the male sex hormones, the androgens. The gene that codes for the androgen receptor (*AR* gene) is found on the X chromosome, so males have only one copy of the gene, while females have two copies. This simple fact makes males more vulnerable to androgen insensitivity syndrome, because a single mutation in the *AR* gene can cause significant loss of function. The receptor simply cannot bind effectively to androgens, so the androgen response element on its target genes is not activated—and nothing changes.

Different mutations cause different degrees of loss of function of the receptor. In the most extreme form, complete androgen insensitivity, the receptor does not respond to androgens at all.

A baby, karyotype 46XY, with complete androgen insensitivity will not be able to override the default situation of developing female genitalia. The baby will produce androgens, but will not be able to respond to them, so all the developmental features that need androgens will not be present. The baby will not have a penis or scrotum but will have testes, although these will be within the abdomen, as there isn't a scrotum. When the baby is born it will be identified as female. As there is no androgen action on the brain, the child will develop a 'female brain' and will identify as fully female. At puberty there will be high levels of androgens and so there is normal breast development because the androgens are converted into oestrogens, needed for breast development. Pubic hair growth needs androgen action so there will not be any pubic hair. Obviously, there can be no menstruation because there aren't any ovaries.

There are less severe forms of androgen insensitivity, where a baby boy is born with an underdeveloped penis or with 'ambiguous genitalia', so that at birth it isn't possible to tell whether the baby is a boy or a girl.

We saw earlier that for testosterone to have most of its effects, it has to be converted into its more potent form dihydrotestosterone (DHT). 5 alpha reductase deficiency is a condition caused by a defect in the enzyme that converts testosterone into DHT. In the developing foetus this enzyme is found in the tubercule, as well as in other tissues. If this enzyme is defective, then the tubercule cannot develop into a penis and the baby is born with female external genitalia. This is an extremely rare condition in most of the world, except in one part of Colombia, where it is quite common. Do some research and find out how these boys develop through puberty.

All of these rare situations throw light on what happens during the development of most foetuses, as well as those affected by the conditions—and in each case they demonstrate the default female form of the body.

3.2   Sex development in other species

Although we have just described situations where sex development goes wrong, sex development in mammals is actually relatively straightforward. If everything develops to plan, females are XX and males are XY. Birds, like mammals, have a pair of sex chromosomes, but these are labelled Z and W, with males being homogametic with ZZ chromosomes and females heterogametic with a ZW pair.

Reptiles have a really extraordinary system of determining sex. In some species the sex is determined by genes and the embryo has a pair of sex chromosomes. Some reptiles use the same system as mammals, and have X and Y chromosomes. Females are homogametic, with XX chromosomes, while males are heterogametic, with XY chromosomes. However, other reptile species use the Z and W form like the birds. So far this is reasonably straightforward. However, not all reptiles have well-defined sex chromosomes and the sex chromosomes present in the embryo don't always match the sex of the baby reptile when it hatches. How can that be? Well, reptiles have a temperature-sensitive sex determination mechanism which can override the genotype determined sex. Let's look at an example.

Most reptiles lay eggs in a nest, which is buried under sand or vegetation. The eggs are often completely abandoned in the nest and hatch some time later. The sex of baby alligators is determined by the temperature of the nest where the eggs are developing, or even the temperature of the eggs within the nest (see Figure 3.2). A cold nest produces more female hatchlings and a hot nest also results in females hatching. But if the egg is just in the middle of the temperature range, then males hatch. The temperature-dependent genes appear to control the aromatase enzyme, responsible for converting male androgens to female oestrogens. More aromatase activity gives more females.

In some reptiles with the ZZ/ZW sex chromosome system, such as the Australian dragon lizard, temperature can override the basic rule that males are ZZ and females are ZW. At a high temperature the ZZ eggs develop into female lizards, but at a lower temperature they develop into males. So female lizards can either be ZZ or ZW, but males are always ZZ.

In fish a similar mechanism is in place, with environmental factors overriding the sex chromosomes. These environmental factors include

Figure 3.2 In reptiles, the sex of the babies is determined partly by their genotype—and partly by the temperature of the eggs as they incubate

rookiephoto19/Shutterstock.com

temperature and social cues. Unlike reptiles, fish can hatch as one gender and develop into the other, or they can hatch in an undifferentiated form and only undergo full sexual differentiation if the conditions are right. In each case the change in environmental conditions causes a change in gene expression, altering the production of sex hormones.

Case study 3.1
Little Nemo's parents—an example of a transgender family

Traditionally people thought of gender as something that usually doesn't change, although this model is becoming more flexible. Most people are born a particular gender, identify with that gender, and live their entire life as a person of that gender. Some people are born a particular gender but never identify with that gender and may choose to undergo gender reassignment. Others identify as no particular gender. But for the majority of human history, for the majority of people, biological sex and the gender we identify with are the same. In some parts of the animal kingdom gender is much more changeable. Let's look at clownfish (*Amphiprion ocellaris*), the type of fish that starred in the Disney Pixar film, *Finding Nemo* (see Figure A).

Figure A Image of clownfish. In real life, when little Nemo's mother died, his father would have changed sex.

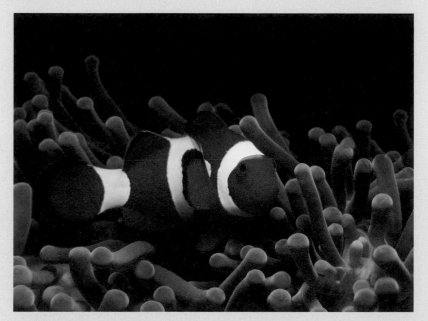

stockpix4u/Shutterstock.com

Clownfish and some other species of fish that live on the coral reef are hermaphrodites, which means that they have both male and female sex organs. However, unlike earthworms and slugs, the two genders are not seen at the same time. Instead these fish have a really interesting development process, starting out as male and eventually becoming female. This is called *sequentially hermaphrodite*. Hormones are, of course, involved in the process of sex change, but social hierarchy is important as well.

Each clownfish family group consists of a breeding female, which is the largest fish in the family, a slightly smaller breeding male, and a number of other, smaller fish which are all male. They live very close to clusters of anemones which give protection from predators, but which also means that each family group can be socially isolated from other families. So within each family the large, dominant female clownfish breeds with the largest male (clownfish are monogamous, meaning that they only have one mate) and produces eggs. The eggs hatch into little male clownfish. Where it gets really interesting is what happens when the female dies. Immediately, the largest male takes over the role of the female, becoming the dominant member of the family group. Over the next few weeks he undergoes a complete sex change, from male to female and he becomes the dominant breeding female. The next largest fish in the family develops into a sexually reproductive male and mates with the new female. In clownfish families, size definitely matters.

At the start of the film, *Finding Nemo*, we see that Nemo's mother (Coral) and most of her eggs are eaten by a barracuda, leaving just Nemo's dad, Marlin, and the single egg that hatches into Nemo. In real life, Marlin's wife, Coral, would have been the head of a large family of clownfish, supported by Marlin, her husband. And as soon as Coral was eaten by the barracuda, Marlin would have started to change. By the time Nemo was ready for school, Marlin would have fully transitioned into a female fish and another fish from the family would have developed into a male to take over Marlin's role as father of the family. It would have made a different—but still very interesting—film!

❓ Pause for thought

1. The fact that the dominant male clownfish changes its behaviour to female the moment the dominant female dies, suggests she may have been sending a signal that suppressed female behaviour in the male. How many ways can you think of for how this signal could be sent?

2. If our hypothesis is that in life a dominant female sends a signal to suppress female behaviour and phenotype in the male, how would you set up experiments to test this hypothesis for all the ideas you had in 1. (assuming you have access to an aquarium)?

3.3   Thyroid hormone in development

We are most familiar with thyroid hormone as a regulator of metabolism and the rate at which our body uses oxygen in respiration. Thyroid hormone also has an important role in development, both in the human body and in other animals. For example, it plays a critical role in amphibian metamorphosis.

Metamorphosis

Of all the vertebrates only amphibians (frogs, toads, and newts) undergo metamorphosis, where the immature form of the animal looks completely different to the adult. Female amphibians lay eggs in water. These are fertilized by males, who climb onto females waiting for them to lay the eggs. The eggs develop in the water, hatching into tadpoles. The tadpoles resemble a rather squat fish, with a round body and a long tail, but they have external gills and no fins or other limbs.

Over a period of several weeks these tadpoles undergo metamorphosis: a process of transformation into the adult frog or toad. This is an extraordinary process, involving remodelling of every tissue type within the body. Muscle, skin, bone, liver, the immune system, the blood system, the reproductive system, the respiratory system, the nervous system; every tissue type undergoes a profound and dramatic change. All of these changes are controlled by thyroid hormone, and they are illustrated in Figure 3.3.

If the tadpole isn't able to get enough iodine to make thyroid hormone, then it gets stuck in the tadpole form. It will continue to grow in size and may live for as long as two years, but it will always remain a large tadpole and never become a frog.

Thyroid hormone acts on the thyroid hormone receptor, one of the nuclear receptors which acts as a transcription factor, and so directly alters gene expression. It isn't too difficult to imagine how this mechanism of action might result in the development of different cell types, changing the epithelial cells of the skin into a different form, for example. What is much more difficult to understand is how thyroid hormone causes a single type of cell to do different things, depending on where the cell is located. This is most obvious when we look at muscle; thyroid hormone causes muscle growth in the new limbs, but causes the muscle in the tadpole tail to disappear.

Figure 3.3 Who would guess these are all the same animal? No wonder a complex interaction of genes, hormones, and enzymes is needed to bring about the transformation.

Nature Photographers Ltd / Alamy Stock Photo

Let's look at the stages of metamorphosis. As the tadpole grows it starts to produce thyroid hormones, only a small amount at first but then in increasing concentrations. The first visible sign of change is the appearance of limb buds at the base of the tail. The limb buds are the first cells to respond to thyroid hormone and only need low levels to start dividing. Gradually, more tissues start to develop as thyroid hormone levels increase. Perhaps the most remarkable change happens at the peak of metamorphosis when, over a period of just five days, the tadpole intestine shortens to just a quarter of its original length. The final tissue to change is the tadpole tail, which disappears. It doesn't drop off, which you might think would be a simple solution, but instead the cells of the muscle and skin in the tail undergo programmed cell death: apoptosis. This tail remodelling needs the highest levels of thyroid hormone and so occurs right at the end of metamorphosis when thyroid hormone levels are highest.

Scientists working in this field, using molecular techniques, have shown that the changes in metamorphosis are not just caused by increased production of thyroid hormone by the tadpole. In addition to the increased production of hormone, there is an increased number of thyroid hormone receptors and an increase in the concentration of the deiodinase enzyme that converts T4 to the much more active T3. So there is a coordinated process involving increased production, activation, and action of thyroid hormone in order to bring about the transformation of a tadpole into a frog.

Thyroid hormone in human development

In Chapter 2 we saw how iodine deficiency usually causes growth of the thyroid gland, resulting in a goitre. Two billion people globally are estimated to be at risk of iodine deficiency. This is more than one-quarter of the world's population, and the map in Figure 3.4 shows you the areas with the biggest problems. As iodine is a key component of thyroid hormones, an iodine deficiency results directly in a lack of thyroid hormone, a disorder which is called hypothyroidism. Iodine is a trace element, present in rainwater and in the soil. People who live on land with low iodine levels are usually iodine deficient because the livestock and plants raised on that soil will also be iodine deficient. If you don't get sufficient iodine, then the thyroid gland in the neck grows larger in an attempt to capture more iodine and make more thyroid hormone. People who live in areas where the water is low in iodine very often have a swollen neck, called a goitre, caused by a greatly enlarged thyroid gland.

Globally, many need some form of dietary iodine supplementation so that they can make sufficient thyroid hormone. Most commonly this supplementation is in table salt, which is enriched with iodine in many countries of the world. So why is iodine, and therefore thyroid hormone, so important that governments around the world act to ensure that their populations receive enough of this trace element? The answer is that the most common preventable cause of brain damage, across the world, is lack of iodine in the diet.

As a baby develops before birth it relies on hormones for growth and development. We have seen how testosterone is critical for the male reproductive system to develop. In a similar way thyroid hormone is essential for

Figure 3.4 The grey areas indicate places where there is no iodine deficiency. The yellow areas indicate regions where few people die from iodine deficiency, orange indicates a higher level of iodine deficiency and so mortalitty, and the red and dark red regions indicate both the highest levels oif iodine deficiency and the highest rates of death as a result.

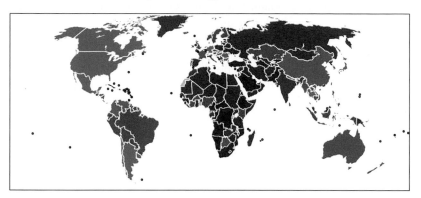

Data from World Health Organization Disease Burden Estimates for 2000-2012. © Chris 55 / Wikimedia Commons / CC BY-SA 4.0

normal foetal development, particularly of the brain. If a mother does not get enough iodine when she is pregnant, the baby is born with a form of severely stunted physical and mental growth, called cretinism. In less severe forms of hypothyroidism in mothers the baby is born with impaired brain function, even if physical growth is normal. For this reason all developed countries in the world test all newborn babies for thyroid function. Even after birth the brain relies heavily on thyroid hormones to continue developing. If a newborn baby is found to have too little thyroid hormone, then it can be given oral thyroid hormone supplements to support growth and development and keep the child healthy.

The importance of iodine in healthy development was recognized early in the twentieth century and by 1920 several countries, including the USA and parts of Europe, had started adding iodide to table salt. However, this practice was slow to spread to the rest of the world. The World Health Organization has launched a campaign for global iodization of salt and, as a result, they report that the incidence of iodine deficiency has halved, with 66% of households now having access to iodized salt and only fifty-four countries still iodine deficient.

Because the brain damage caused by untreated hypothyroidism is preventable, every newborn baby in developed countries is tested for hypothyroidism. Two or three days after birth every baby is checked by a health visitor or nurse, who carries out a 'heel prick test'. The baby's heel is pricked to allow collection of a tiny blood sample onto absorbent paper. The blood sample is then checked for a range of disorders which could affect the development of the baby. As thyroid function is so important for development of the brain and nervous system, the heel prick test includes a check for thyroid function. The test doesn't measure thyroxine or T3 directly though. Instead it measures thyroid stimulating hormone. Why do you think that might be?

3.4   Insect development

It may not surprise you to learn that insects have a rather different set of hormones to humans and non-human vertebrates. Their lifecycle is rather different as well, so, although frogs and also many insects undergo metamorphosis, it is a very different process, controlled by very different sets of regulatory hormones, as you can see in Figures 3.5 and 3.6. It's not just insects but also the other arthropods, the crustaceans (crabs, shrimp, and woodlice), that have a distinctly different larval and adult stage. These invertebrates do not use thyroid hormones to control their metamorphosis, but instead have their own set of hormones: ecdysteroids, juvenile hormones, and a great number of neuropeptides.

Figure 3.5 Insect metamorphosis is just as dramatic as that of amphibians—who would have thought that a caterpillar could turn into a butterfly or moth?

Anthony Short

Figure 3.6 The endocrine system in insects looks very different from that of vertebrates. The prothoracic gland (PTG) makes ecdysteroids; the corpus allatum (CA) makes juvenile hormones; and the neurosecretory cells (NSC) make a range of neuropeptides.

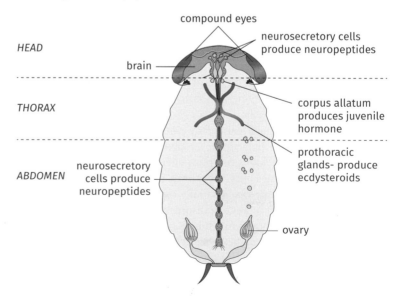

Figure 3.7 The ecdysones are similar to mammalian steroid hormones in structure—but the insect juvenile hormones are in a class of their own

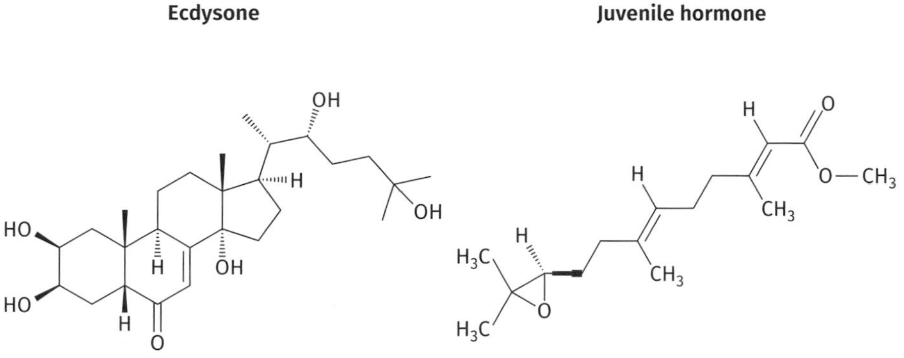

Ecdysone   **Juvenile hormone**

Ecdysteroids have the same basic structure as mammalian steroid hormones, but with significant differences. Juvenile hormones are a family of hormones which vary between different species but are all chemically similar and have a very different structure to any vertebrate hormone. You can see these chemical structures in Figure 3.7.

In the larval stage, insects go through several cycles of growing, moulting, and growing again—see Figure 3.8. The moulting is a process of shedding the outer layer, the hard exoskeleton, so that the larva can grow larger. Typically, insect larvae go through five moulting cycles, with the

Figure 3.8 The lifecycle of a butterfly showing complete metamorphosis. As it grows, the larva moults several times before forming a pupa as it metamorphoses into a butterfly.

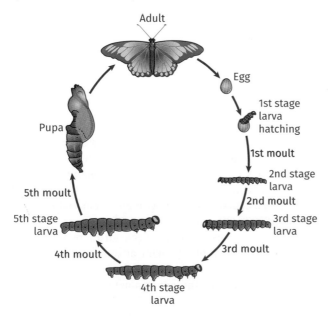

final cycle including metamorphosis into the adult form. The processes of moulting and metamorphosis are controlled by hormones. When an insect larva reaches a certain size, the brain signals to the prothoracic gland to produce ecdysteroids. These steroids cause the old cuticle to loosen and fall off and the new cuticle to develop underneath. Juvenile hormone, produced by the corpus allatum, stops the larva from going through metamorphosis for the first four moult cycles. However, in the fifth cycle, when the insect larva is big enough, the brain tells the corpus allatum to stop producing juvenile hormone, while there is a surge in ecdysteroid production. This allows the larva to finally form a pupa (an inactive or 'resting' stage of development) and undergo metamorphosis, and for the adult version of the insect to emerge. So the role of ecdysteroids is to make the insect moult and the role of juvenile hormone is to stop the insect larva from going through metamorphosis until it has grown large enough to be a successful adult insect.

You might think that, once the adult form of the insect has emerged, juvenile hormone would have no further actions. It still remains important in the adult insect, however, as it is essential for different aspects of reproduction. Juvenile hormone is necessary in the process of egg formation and without it insects cannot produce viable eggs.

3.5   Endocrine disrupting chemicals

Many natural and man-made products contain chemicals which are similar to the hormones that our bodies produce. If we are exposed to high enough levels of these chemicals, in food or drinking water, for example, they can interfere with our endocrine system and so are called endocrine-disrupting chemicals. These chemicals can have their effects by mimicking the effects of our hormones or they can exert anti-hormone effects. The effects that have been demonstrated include:

- interference with reproduction;
- developmental malformations;
- increased cancer risk; and
- disturbances in the immune and nervous system function.

Some of the chemicals that are known to have endocrine-disrupting effects are medicines that were designed to have exactly that effect; the contraceptive pill is a good example. Others are natural plant products such as the oestrogen-like steroids (phytoestrogens) found in many plants. Of greater concern are those chemicals where the endocrine-disrupting effect is an unwanted action, particularly as we now realize that endocrine disruptors cause significant damage to wildlife.

Over the past thirty years there has been an increasing awareness of the impact of endocrine-disrupting chemicals on wildlife populations and potentially on human health. The particular concern is about man-made chemicals that are present in the environment, in water, plants, in our food, and in the air. Some of the most widespread and harmful of these chemicals are pesticides that have been used to control insect populations and have turned out to have very wide-ranging harmful effects. Other chemicals are the by-products of manufacturing processes which have entered the food chain.

There is clear evidence that environmental endocrine-disrupting chemicals harm wildlife populations (see Figure 3.9) but there is little information about their effect on human health. Very few chemicals have actually been tested for their endocrine-disrupting effects. Concerns about the impact of chemicals on the environment go back to 1962, when Rachel Carson wrote an influential book called *Silent Spring*. She argued that the indiscriminate spraying of pesticides, such as DDT, caused long-lasting and highly damaging effects on the environment and on wildlife populations. It was largely as a result of her work that DDT was eventually banned across the world. We now know that DDT, its breakdown products, and several closely related chemicals, have many serious effects on wildlife, including endocrine effects. Case study 3.2 gives a specific example.

Figure 3.9 As top predators, polar bears are particularly vulnerable to the effects of endocrine-disrupting chemicals, which accumulate in fat stores of many animals, including the seals that these bears eat. The development of polar bears has been affected by these chemicals, leading to disrupted sexual differentiation.

Gudkov andrey/Shutterstock.com

Case study 3.2

Gender-bending chemicals

Example 1: Apopka alligators

Working in the 1980s, at Lake Apopka in Florida, an environmental scientist called Dr Louis Guillette noticed that there was a problem with the local alligator population. There were fewer eggs hatching and the young alligators were less likely to survive than in other populations. He also saw that the penises of male alligators were getting smaller and the female alligators had abnormal ovaries. He linked these findings to an accident in 1980 at the nearby Tower Chemical Company, which had resulted in a big spill of a pesticide, dicofol, into the lake. Dicofol is very closely related to DDT but still legally used in many countries around the world as a pesticide to control mite infestations of poultry and livestock. Dr Guillette found that the Lake Apopka alligators had very high levels of dicofol breakdown products in their blood and suggested a link between the pesticide and the damaged reproductive functions of the alligators.

It was this work that led to the concept of 'gender-bender' chemicals: endocrine disruptors which affected the sexual development of animals. Since then, there have been many examples of such chemicals and animals affected by them. The 'gender-bender' endocrine disruptors act by either having the same effects as oestrogens, the female sex hormones, or acting as anti-androgens, blocking the production or actions of male sex hormones. These chemicals can come from prescription medicines, from manufacturing processes, and from pesticides. In the environment these endocrine disruptors can also come from sewage effluent which contains oestrogens naturally excreted by women.

Example 2: Diethylstilboestrol defects

Between 1940 and 1975, several million pregnant women in the USA and Europe were prescribed a synthetic oestrogen-like drug called diethylstilboestrol, or DES. The idea was that DES would prevent miscarriage or premature birth. Unfortunately, it didn't work. Even worse, it also caused significant health problems in the children born from these pregnancies. The simple fact that DES was used for such a long time shows that it didn't cause problems that were obvious at the time, but as these children grew into adults several health issues became clear. The daughters who had been exposed to DES were forty times more likely to develop a rare form of vaginal cancer. They were also much more likely to have problems with conceiving and with pregnancies, including an increased risk of miscarriage. The men who had been exposed to DES were almost three times more likely to be born with an undescended testis, and as adults were much more likely to develop problems with the testes, including inflammation and testicular cancer. More recent research

Figure A Birth defects affecting the genitals such as the hypospadias shown here are becoming more and more common—is there a link to the everyday chemicals we use?

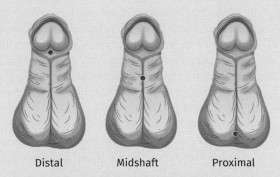

Distal   Midshaft   Proximal

● Denotes position of external urethral meatus—
the opening through which urine and semen pass

© Mayo Foundation for Medical Education and Research.

has shown that the problems continue into the next generation. The sons of women who were exposed to DES before birth are much more likely to have developmental problems, particularly **hypospadias** (see Figure A) and undescended testes (**cryptorchidism**).

❓ Pause for thought

1 Many of the pesticides used in agriculture are designed to stop crop damage by insects. One of the problems of many pesticides is that they don't only affect the species that is causing the problem, but kill other species as well, especially at the top of the food chain where bioaccumulation of pesticides is an issue. Could you use our knowledge of insect hormones to design a pesticide that would only affect the insect pest? Would you target ecdysteroids or juvenile hormone?

2 How do you think that the effects of DES can affect the grandchildren of the women who took this drug? What mechanism could explain this effect?

Pesticide problems

We need pesticides. The world population is growing all the time and pests destroy millions of tonnes of crops, which are needed to keep people alive. The problem is that the chemicals which control pests can have far-reaching and unexpected effects. Modern research into new pesticides involves testing to the levels required for new human drugs, and includes testing for the environmental impact (see Paul Beales et al., *Plant Diseases and Biosecurity* in this series). However, historically we simply did not have this level of awareness—and some extremely dangerous chemicals have been used with far-reaching effects.

Both dicofol and DDT belong to a group of pesticides called organo-chlorides. Many of these chemicals are endocrine disruptors, including the highly controversial insecticide, endosulfan. Endosulfan has been widely used around the world, sprayed onto crops of cotton, tomatoes, potatoes, and other foodstuffs, to kill insects which damage the crop. It was designed to be a contact insecticide, killing insects that came into direct contact with it. More than 600,000 kg of endosulfan were used every year in the USA between 1987 and 1997. This chemical is described as a 'ubiquitous environmental contaminant' and has been found in locations many thousands of miles away from where it has been used. Endosulfan is found in the Arctic Ocean, in the air above the Antarctic, in sand from the Sahara desert, and in water, plants, and animals all around the world. It breaks down very slowly and so, during its period of use, has gradually accumulated in the biosphere. It is considered to be the most abundant pesticide in the world. When humans and other animals are exposed to high levels of endosulfan it is a nerve poison, disrupting nerve transmission, but in the environment it is present at much lower concentrations. So why does it concern us?

Endosulfan is a chemical that mimics oestrogens, the female sex steroids. It binds to oestrogen receptors and has effects similar to the effects of oestrogens. It also inhibits the aromatase enzyme, which converts androgens to oestrogens in the body. In wildlife populations endosulfan interferes with the effects of both ecdysone and juvenile hormone in crustaceans, affecting growth and sexual development. It disrupts breeding cycles of many different types of beneficial insects—for example the bees and hoverflies we need to pollinate many food plants—and it decreases fertility in fish populations and in birds. It disrupts both male and female sex hormones in all types of vertebrates, decreasing fertility and increasing the number of birth defects affecting the reproductive system. It also disrupts thyroid hormone production, affecting metamorphosis in amphibians.

Exposure to endosulfan also damages human sexual development. Cashew nuts are an important crop in the Kasargod district of the Indian state of Kerala. For over twenty years the cashew plantations were treated with endosulfan and no other pesticide. As a result the local environment in Kasargod was completely contaminated with endosulfan. Researchers compared the development of boys in this district with boys in a similar district that did not use endosulfan; this is probably the best *model system* you could get for looking at the effects of a single pesticide. The researchers found that compared with the boys in an unpolluted region, the boys in Kasargod had measurable levels of endosulfan in their blood, lower levels of testosterone, delayed puberty, and a higher incidence of male reproductive birth defects (hypospadias and cryptorchidism).

This was sufficient evidence for the Keralan authorities to ban the use of endosulfan as a pesticide. Many other countries have followed and the use of endosulfan is gradually being phased out. Scandalously, some countries, which have prohibited its use on their own farms, still export it to other countries which have not yet banned this damaging poison.

Scientific approach 3.1

How do we investigate the effect of pesticides on endocrine systems?

The effects of pesticides on endocrine function have been discovered by working with experimental animals, mostly rats and mice, and with cultured cell lines or tissues in the laboratory. The effects of endosulfan on aromatase activity were investigated using microsomes from human placental tissue. The placenta, which is expelled from the mother's body after a baby is born, can be a really useful research tool. The cellular organelles called microsomes are the location of the enzyme aromatase. The placenta has a very high concentration of aromatase, so placental microsomes are ideal for use in aromatase assays: measuring the level of aromatase activity. A standard preparation is made of microsomes and aromatase activity is measured in the presence of increasing concentrations of the pesticide, to determine whether it affects aromatase activity and at what concentrations.

Immortalized cell lines are also often used in this type of research. Many cancers are a form of immortalized cell: the cancer cells grow outside the normal cell growth regulation mechanisms. This can mean that, when the tumour cells are grown in a special nutrient fluid, the cells divide but always keep the same characteristics, becoming an *immortalized cell line*. There are cell lines from many different cell types available to researchers, including some cells that respond to oestrogens by proliferating, which means growing and dividing more rapidly. The effects of endosulfan have been investigated on a cell line called MCF7 cells. These cells proliferate when exposed to oestrogens and researchers found that this effect is increased by endosulfan.

So these laboratory techniques, called **in vitro methods** (literally *in glass* methods) can tell us a lot about the biochemical and molecular mechanisms that pesticides and other chemicals might affect, but they don't tell us what happens to whole organisms, wildlife populations, or people. Many chemicals go through a process of animal testing before they are used, so that we know whether they are toxic and what adverse effects they might have. These tests are used to establish the lethal dose of a chemical, so we know whether it is likely to kill you if you accidentally swallow some, but don't always look at the effects of being exposed to low concentrations of the chemical over a longer time. These effects are likely to be more subtle and very different between different species. For example, the version of aromatase produced in humans may be more sensitive to endosulfan than the version of aromatase produced by a rat.

For ethical reasons, it is extremely difficult to design experiments to test environmental endocrine disruptors on human populations or on wildlife populations and so most of our information in these areas comes from accounts of accidental exposure, such as the effects of dicofol on the Lake Apopka alligators and the effects of endosulfan on the sexual development of boys in Kasargod, Kerala.

? Pause for thought

It is very unusual for a population to be exposed to just a single pesticide. How can we determine the effects of low-level exposure to multiple chemicals over a long time period?

Chemicals from manufacturing

We all use plastic containers for water and soft drinks, for collecting takeaway food, and for food storage. We also use plastics to wrap food: supermarket pizzas; fruit and vegetables; cheese; bread. Everything comes wrapped in plastic. Even snacks like crisps, chocolate, and other sorts of sweets all come in plastic wrappers. This is a recent change in human habits. These types of flexible, transparent, plastic containers and wrappers have only become available over the past fifty or sixty years. Before that we used paper, metal, or glass containers for our food. Apart from the extraordinary amount of waste created by single-use plastic containers, should we worry? We know about the harmful physical effects that plastics have, particularly on marine ecosystems, but there is a lot of evidence that the chemicals involved in plastic manufacture and within the plastics also have damaging effects on the endocrine system.

The manufacture of plastics involves the use of chemicals which act as oestrogens (see Figure 3.10). Some of these chemicals remain in the final plastic product and there has been a lot of concern about whether they can get into foodstuffs that are wrapped or stored in plastic containers. Bisphenol A (BPA) is very widely used to make plastics for food storage, including the linings of many food and drink cans, and is also used in children's toys, in printer ink, and in paper recycling processes. Its use as a flame retardant in foam and furniture means that it is one of the most widely used chemicals. Phthalates are used to make plastics more bendy and less brittle, so have been very widely used in food wrappers. We now know that both these chemicals can be called 'gender benders'. BPA has oestrogenic actions and phthalates act as anti-androgens. Other chemicals used to make plastics are known to disrupt thyroid function and we have seen how that can affect the development of humans and other species.

Studies using experimental animals have shown that phthalates block androgen production by the testes, rather than blocking androgen action at its receptors. The result is developmental abnormalities which look very similar to the effects of androgen resistance in humans (see section 3.1). There are also malformations in the testes and a greatly reduced sperm count. Bisphenol A has accumulated in landfill sites and has contaminated local water supplies. It has affected fish populations and, from there, up the food chain; and it also seems to be affecting the people and animals who drink the water. In the USA it has been found that over 90% of people have measurable levels of

Figure 3.10 All of these compounds can bind to the oestrogen receptor. What structural similarities can you see that allows these very different compounds to all bind to the same receptor?

17B oestradiol—the major naturally occurring oestrogen

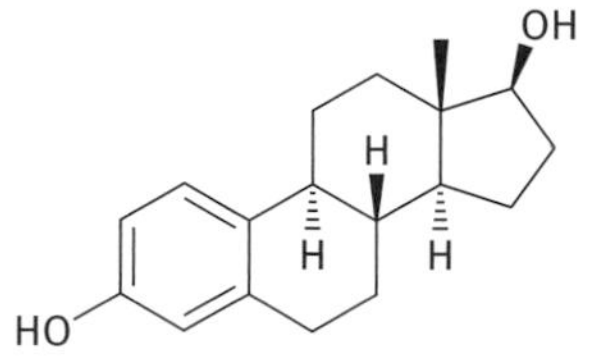

Tamoxifen—a selective oestrogen receptor modulator

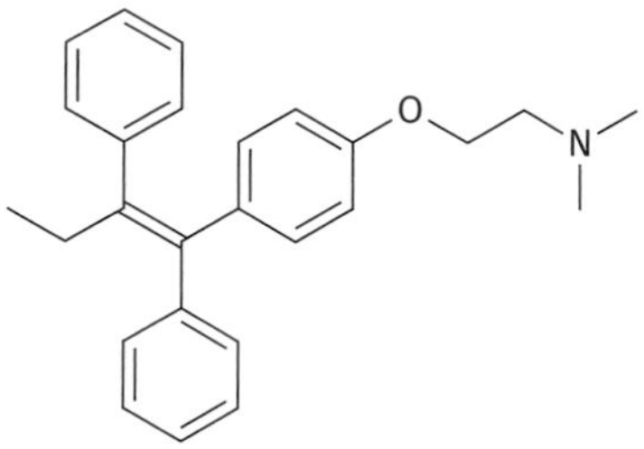

Genistein—a plant steroid

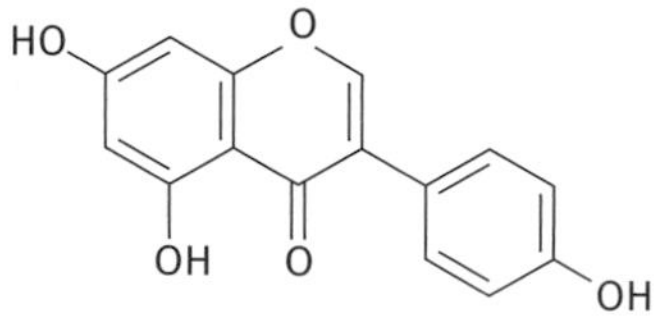

Bisphenol A—an industrial chemical

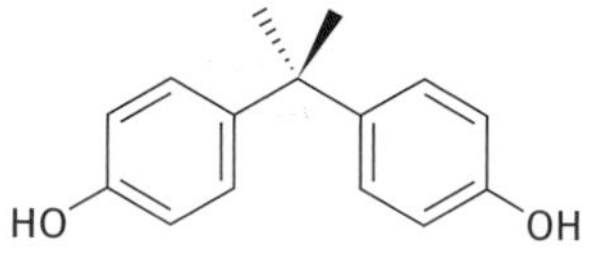

Diethystilboestrol (DES)—a synthetic oestrogen

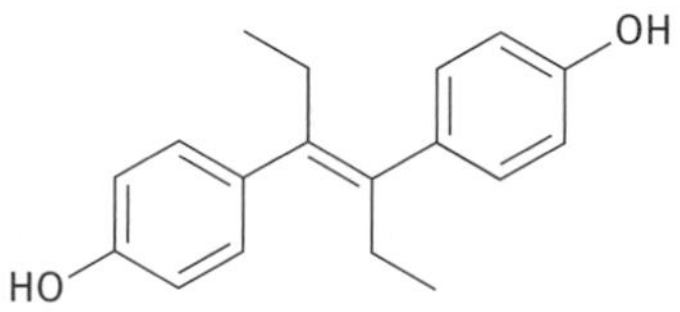

Ethinyloestradiol—the main oestrogen used in the contraceptive pill

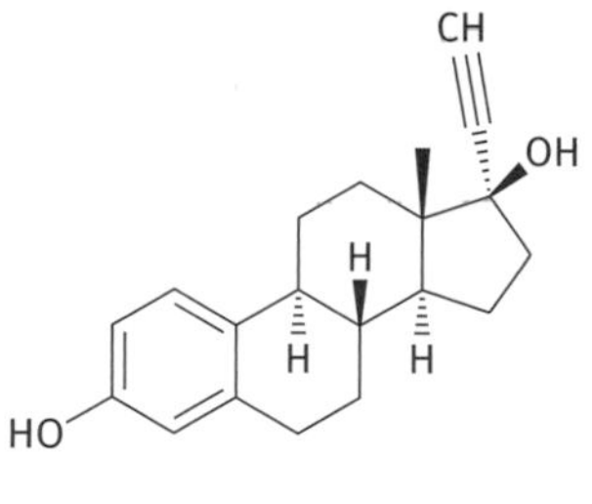

BPA in their blood. BPA has caused a decrease in fertility of minnows and other fish, reducing both egg and sperm production. It has the same action in trout and in all other fish species investigated. In amphibians BPA causes a huge decrease in the numbers of mature males, shifting the development of tadpoles towards becoming female. In reptiles, such as turtles, it has the same effect, reducing the numbers of males in the population. In all these vertebrates BPA causes developmental disorders, particularly of the male sex organs. As we move up the food chain, looking at predator species, it becomes more difficult to judge whether one specific agent is responsible for developmental problems. We do know that in mammals that feed on fish (e.g. otters, seals), there has been a decrease in fertility and an increase in the numbers of males with small or abnormal penises.

Plant steroids

Many plants produce steroids that are similar to oestrogens. These are called **phytoestrogens**. The main plant oestrogen is called genistein. Its structure is shown in Figure 3.10. There is little evidence that these steroids have any developmental effects on the animals that eat them, although they do affect fertility in some species. We will look much closer at plant steroids in Chapter 6.

Effects on human sexual development

The question of whether endocrine disruptors affect human development is controversial. We know that men are becoming less fertile because they are producing fewer healthy sperm. The number of sperm per millilitre of semen has halved over the past few decades (see Figure 3.11). The number of men with a sperm count so low that it affects their fertility has more than doubled since 1940 and the trend appears to be continuing. The numbers of baby boys born with an incorrectly developed penis (called a hypospadia, see Figure CS 3.2) has also increased dramatically, with twice the number seen now compared to the 1970s. We also know that between 1940 and 1970 in England there was a doubling of the number of boys born with an undescended testis, called cryptorchidism, literally, *hidden testis*. So we know that there is a worldwide problem with the human male reproductive system; what we don't know is what is causing the problem. Is there any evidence that it might be a result of endocrine-disrupting chemicals?

There is no direct evidence and of course it is impossible to carry out experiments to test the theory. We know that huge amounts of pesticides are produced and used worldwide and that the use of plastics has increased greatly over the past sixty years. We have noted how widespread the endocrine-disrupting chemicals have become in the environment, in our food, and in our homes. We know that this is reflected by significant concentrations of some of these chemicals being present in the blood of most humans. Over this time we have seen a huge increase in reproductive problems, particularly in men. These problems exactly mirror the damage to wild-life populations and experimental animals exposed to the same chemicals. So while there is no direct evidence to link endocrine disrupting chemicals with decreasing fertility and an increasing incidence of developmental problems in baby boys, there is a very great deal of circumstantial evidence.

But perhaps we should end our discussion of endocrine disruptors on a slightly more hopeful note. It is now possible to buy food wrapping and plastic food containers that are phthalate-free. Bisphenol A is being replaced in many plastic manufacturing processes, although it is still widely used in other areas. Water and sewage treatment is improving and there are programmes to remove oestrogenic compounds from water. Some of the nastier organochloride pesticides are being phased out, although there is still a long way to go. Plants will always contain phytoestrogens and there is little we can do about that! In Chapter 6 we shall see how these and some other chemicals with endocrine effects can be used to give positive health benefits.

Figure 3.11 Changes in sperm count in different continents, 1975–2020.

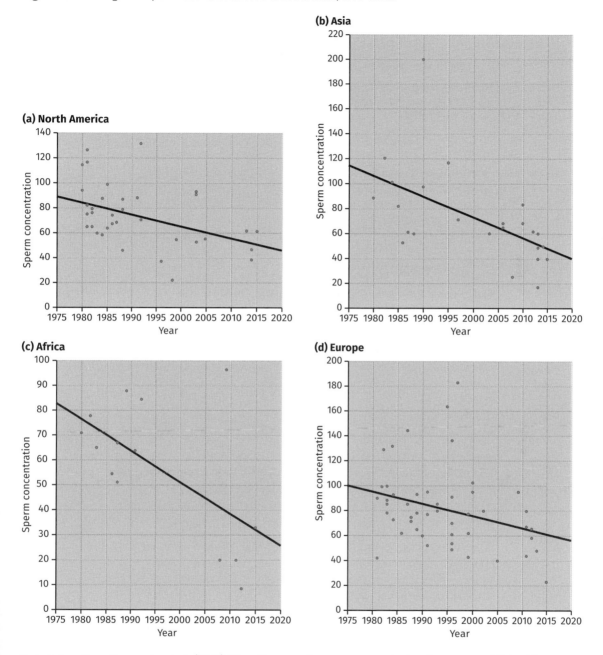

Data taken from Sengupta et al. (2017) 'The disappearing sperms: analysis of reports published between 1980 and 2015', *American Journal of Men's Health* 11, 1279–304.

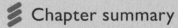

Chapter summary

- In mammals, including humans, hormones are essential for the development of a male phenotype. The female phenotype is the default and it is the presence of male hormones that overrides the default and allows the baby to become phenotypically male.
- Much of our understanding of normal sexual development comes from investigating situations where it goes wrong.
- In other vertebrates, gender determination is more complex and can be affected by factors such as temperature and social hierarchy.
- Hormones control metamorphosis in vertebrates and insects.
- Many natural and man-made products contain chemicals which are similar to hormones and can interfere with normal hormonal processes. It is increasingly recognized that these chemicals have damaging effects.

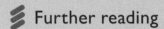

Further reading

Rachel Carson, *Silent Spring* (Penguin Modern Classics)
First published in 1962, this book is credited with first raising the issue of pesticides as an environmental problem.

Wissem Mnif, Aziza Ibn Hadj Hassine, Aicha Bouaziz, Aghleb Bartegi, Olivier Thomas, and Benoit Roig (2011). 'Effect of endocrine disruptor pesticides: a review', *J. Environ. Res. Public Health* 8, 2265–303; doi:10.3390/ijerph8062265.
This freely available journal article contains a lot of information about the extraordinary range of pesticides that have endocrine-disrupting effects.

Muthayya, S., Rah, J.H., Sugimoto, J.D., Roos, F.F., Kraemer, K., et al. (2013). 'The global hidden hunger indices and maps: an advocacy tool for action', *PLoS ONE* 8(6): e67860. doi:10.1371/journal.pone.0067860.
This freely available research paper explores the issues around iodine deficiency on a global scale.

Talsness, C.E., Andrade, A.J., Kuriyama, S.N., Taylor, J.A., and Vom Saal, F.S. (2009). 'Components of plastic: experimental studies in animals and relevance for human health', *Philos. Trans. R. Soc. Lond. B Biol. Sci.* 364(1526), 2079–96. doi: 10.1098/rstb.2008.0281.
This freely available review article describes experiments carried out to look at the effects of plastics on animal health and draws conclusions about human health risks.

Vandenberg, L.N., Colborn, T., Hayes, T.B., Heindel, J.J., Jacobs, D.R., Lee, D-H., Shioda, T., Soto, A.M., vom Saal, F.S., Welshons, W.V., Zoeller, R.T., and Myers, J.P. (2012). 'Hormones and endocrine-disrupting chemicals: low-dose effects and nonmonotonic dose responses', *Endocr. Rev.* 33, 378–455. doi: 10.1210/er.2011-1050. doi:10.1210/er.2011-1050.

https://www.scientificamerican.com/article/experts-temperature-sex-determination-reptiles/

Sharp, R. (2012), 'Sperm counts and fertility in men', *Embo reports* This article reviews the evidence around falling sperm counts, available at https://www.ncbi.nlm.nih.gov/pmc/articles/PMC3343360/pdf/embor201250a.pdf

Todd, E.V., Liu, H., Muncaster, S., and Gemmell, N.J. (2016). 'Bending genders: the biology of natural sex change in fish', *Sex Dev.* 10, 223–41. This freely available scientific report explores the mechanisms influencing sex change in fish.

 ## Discussion questions

3.1 Why is the default mammalian sexual development female? What advantage might this have over a default male development?

3.2 In this chapter we discussed in vitro methods of research, specifically studying immortalized cell lines in the laboratory. In vivo means carrying out research on a living organism, which for our species is a human being. What are the pros and cons of each approach? What is each most suited to and what drawbacks can you imagine?

4 HORMONES AND REPRODUCTION

In Chapter 3 we thought about how young animals have to go through a lot of changes before they are ready to replicate themselves. Now that we have considered development we will look at the actual process of reproduction. Even if we just look at mammals we will find differences between cats, camels, sheep, and ourselves. We should start by making it clear what we're talking about. We will be looking at the testes which are the main male reproductive tissue and the ovaries which are the main female reproductive tissue. Both of these glands have two functions: the production of gametes and the production of steroid hormones. The testes produce sperm and androgens while the ovaries produce ova (eggs) and oestrogens and progesterone (see Figure 4.1).

Figure 4.1 This scanning electron micrograph shows an ovum surrounded by sperm. Mature eggs and sperm are produced after puberty in response to a number of different hormones

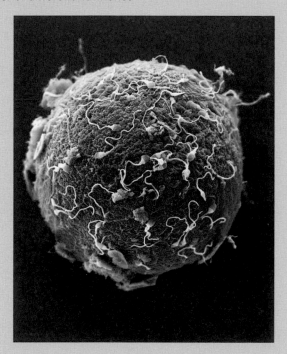

Eye of Science/ Science Photo Library.

4.1   Puberty: the start of our reproductive lives

Your relatives will probably have pointed out to you (every time they see you) that you keep on growing. As we increase in age, from birth to our teens, we also increase in height. The increased height is largely due to growth of our bones, especially the bones in the legs. As our height increases we have a proportionate growth in our skin, muscle, and other tissues. This process of bone and tissue growth is mostly controlled by hormones. You won't be surprised to learn that growth hormone from the pituitary gland plays an important part. But in reality a whole cocktail of hormones is in- volved, including thyroid hormones, steroids from the adrenal gland, and insulin from the pancreas. If any part of the endocrine system isn't working properly then growth is affected. It's probably worth adding that there are many other factors that affect the growth of children, including any kind of illness or emotional trauma, particularly if it lasts for any length of time. In part this effect is because *stress*, whether physical or emotional, affects so many hormones, including growth hormone. The rate at which we grow as children is important because it determines how quickly we enter the next stage of development: puberty.

What triggers puberty?

The transition from childhood to adulthood in biological terms is called puberty. Puberty is the process of development into sexually mature adults—you can see the main stages in Figure 4.2. There is no set age for the start of puberty and it can be difficult to decide exactly when somebody has started this stage of development, particularly in boys, where the physical changes are gradual.

It is a bit easier to identify puberty in girls as there is a single event which can be dated: the first menstrual period. This is called **menarche**. Working from this event, scientists have been able to determine the main factor controlling puberty and the onset of menstruation. It is simply size: when a healthy girl reaches a weight of about 47 kg she enters menarche and begins to menstruate. In boys there does not seem to be quite the same weight cut-off. In fact, the trigger for puberty in boys is more complex and is determined by both body weight and genetics. Studies in America used testicular volume as an indicator of stage of puberty. They found that white boys who were overweight started puberty earlier than boys of normal body weight, while obese boys had delayed puberty. In Hispanic boys, body weight had no impact at all on the timing of puberty. There is still a lot of research that needs to be carried out to understand the trigger for the onset of puberty in boys.

In girls the concept of a threshold weight explains much of the age variation in the time that girls start their periods; age isn't important at all and the only thing that really matters is weight. This also helps explain why girls have been starting their periods at increasingly younger ages over the past one hundred years. Over this time there has been a major change in childhood health. A hundred years ago most people did not have access to a toilet inside their home; now in the developed world almost everybody does. Nutrition has improved so that children grow bigger and more quickly. The increase in availability of vaccinations together with improved nutrition and sanitation means that there are fewer childhood illnesses which, as we have seen, disrupt normal growth. All these factors together mean that girls reach the magic body mass of 47 kg much younger than they did a hundred years ago and so start their periods earlier. In purely biological terms it makes a lot of sense; once an individual has reached a certain size then they are biologically equipped to reproduce. Being human, with our complex and highly developed social structures, we understand that there is a lot more to being ready to have children than simply reaching 47 kg in weight but, biologically at least, it is quite simple.

One question we haven't addressed is what happens in girls at a body weight of 47 kg that didn't happen at 45 kg? In other words, what is the switch that kick-starts menarche? Some aspects of this process are very well understood but other questions are at the cutting edge of endocrine research.

We know that the hormones that control the ovary are not produced during childhood and are produced in a very specific pattern during adulthood. These are the hormones of the hypothalamus and pituitary gland; see Figure 4.3. During puberty the hypothalamus starts to produce a hormone called GnRH (gonadotropin releasing hormone) which acts on the pituitary gland causing it to release LH (luteinizing hormone) and FSH (**follicle stimulating hormone**). These two hormones act on the ovary. Luteinizing hormone

Figure 4.2 Physical changes during puberty, as originally described by Professor James Tanner, a British paediatrician. In both boys and girls, the first sign of puberty is the development of pubic hair. Boys usually start to develop pubic hair between the ages of ten and fourteen, but this can occur earlier in girls.

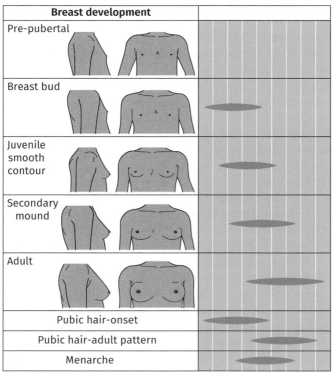

mainly regulates hormone production, while FSH mainly causes the development of a follicle and release of an ovum (egg) every month. In order to do this there has to be a clear pattern of hormone secretion which changes during the month. So the production of GnRH by the hypothalamus is the key step here; without this hormone LH and FSH are not produced and there is no menstrual cycle.

The hypothalamus is part of the brain which works like a central control unit, receiving all kinds of messages from within our bodies and outside and putting out an appropriate neural and hormonal response. It makes perfect sense for this to be the place that initiates the signal for the start of menstruation. Let's think about what signals it could receive when the magic number of 47 kg is reached. At this body mass the body contains a certain amount of fat, properly called adipose tissue. Adipocytes (the cells that make up adipose tissue) make hormones; they make a peptide hormone called leptin which acts on the hypothalamus to regulate our appetite. Adipocytes also make oestrogens, the female sex hormones, by converting adrenal steroids. Experiments on mice have shown that, if the gene for leptin is knocked out, then the mice don't reach sexual maturity. As far as we currently understand, it seems that leptin and oestrogens both act on the hypothalamus to cause the development of a specific group of neurons. These neurons use a novel peptide neurotransmitter, which has been called kisspeptin. Unfortunately, this is not a hormone that improves your kissing. In fact it is a rather cheesy reference to the town where the gene coding kisspeptin was discovered; it is the home of a type of chocolate called 'Kisses'—scientists are humans too!

Kisspeptin neurons coordinate the activity of the hypothalamus, which is the key regulator of sex hormone production. These neurons stimulate the hypothalamus to produce the hormone cascade needed for the ovaries and testes to start producing hormones and gametes—Figure 4.3 shows you the whole process.

The maturation of the kisspeptin neurons changes the pattern of GnRH production over a relatively long time period. At first, before any signs of puberty, there are small pulses of GnRH which happen at night. As the pulses get larger and more frequent they happen during the day as well. Eventually GnRH production settles into a ninety minute rhythm, with pulses during the day and at night. This establishment of a distinct pattern of GnRH secretion over a period of time causes a gradual increase in production of pituitary hormones and the activation of hormone production in the ovaries and testes. The same process happens in boys as in girls. The kisspeptin neurons set the timing for the episodic secretion of the hypothalamic hormone GnRH, which is critical for its action.

If GnRH is produced and released continuously, then it is completely ineffective at stimulating the pituitary gland. It needs to be released in pulses of a certain frequency and amplitude. Kisspeptin neurons generate these pulses. Exactly the same process of hypothalamic maturation happens in boys as in girls, although we don't fully understand the triggers for this process in boys.

Activation of kisspeptin neurons and GnRH pulses is the key to ovulation and **spermatogenesis** in all mammalian species. It gets really interesting when we start to think about the monthly variation in hormone production in women, necessary for ovulation, which simply doesn't happen in boys because sperm production requires only fairly constant FSH and

Figure 4.3 The hormonal control of gametogenesis. Kisspeptin neurones have pulsatile activity, which sets the pattern for GnRH secretion. GnRH, released from the hypothalamus, travels in a local circulatory system to the pituitary and stimulates the release of LH and FSH. In the gonad there is a dotted line from steroid production to gamete production; this pathway is important in the testes but not in the ovary.

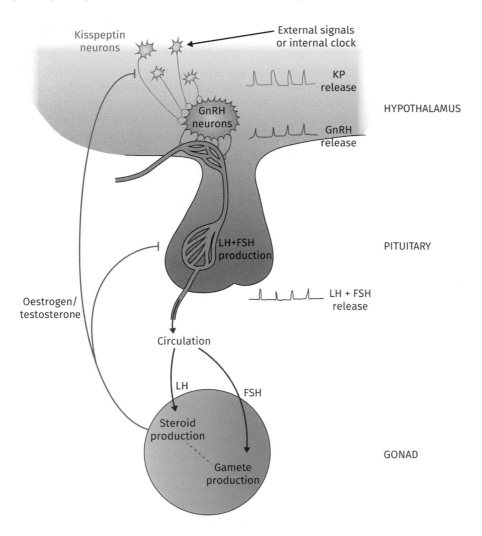

testosterone levels. The patterns of sexual activity and ovulation vary greatly between different species and we will explore three different ways of controlling reproduction: spontaneous ovulation, seasonal ovulation, and induced ovulation.

4.2 Ovulation: the female side of reproduction

Female animals need to produce an egg (ovum) to be able to reproduce and pass their genes on to the next generation. When a female mammal is born, whether she is a female human, kitten, or lamb, her ovaries contain all the eggs that she will ever produce. Baby girls have between one and two million eggs in their ovaries. Only a tiny proportion of these eggs, no more than about 500, will ever be released in the menstrual cycle—the rest just gradually die off. All the eggs start in an immature form and need to go through a process of maturation, when they complete their meiotic cell division, before they are released from the ovary. This process of maturation and egg release is called **ovulation** and it is controlled by hormones. Different animals have evolved very different patterns of ovulation: women usually release only one egg at a time and ovulate every month; rats and mice can release eight or ten eggs at once and ovulate every four days; sheep and goats release one or two eggs at once but only do this once or twice each year; and cats and camels only ovulate after having sex. We will look at how hormones work to control these very different approaches.

Spontaneous ovulation

Some animals are sexually active only at certain times of the year. Others don't have cycles at all, but ovulate in response to the presence of males or the act of sexual intercourse. We will look at each of these, starting with spontaneous ovulation and the human ovarian cycle.

Continuous cyclers: women, rats, and dogs

There are only a very few species that have a female reproductive cycle like ours, with a regular shedding of the uterine lining and bleeding. Most species that have a regular cycle, such as rats, which have a four-day cycle, and dogs, which have a cycle only about twice per year, don't actually shed the uterine lining, but reabsorb it.

From the time of menarche, at age ten to thirteen, right up until menopause, at the age of forty-nine to fifty-one, women produce an ovum roughly every twenty-eight days. This regular ovulation usually stops only if we are pregnant, using a hormonal contraceptive, or in response to some illness or weight loss.

The monthly cycle in women happens because every twenty-eight days or so there is a *follicle* selected in one of the ovaries, which grows, bursts, and heals. It is this repeating pattern that causes hormonal changes (see Figure 4.4). A follicle contains the ovum, surrounded by a layer of cells which produce hormones, mainly oestrogens. Follicle stimulating hormone (FSH) is named because it makes the follicle grow and mature. Luteinizing hormone (LH) controls the oestrogen production by the follicle and LH production is negatively regulated, like most endocrine systems. High levels of oestrogens inhibit LH production and so limit the *drive* to produce more oestrogens. The oestrogens released by the follicle act on the lining of the uterus, making it grow and develop a rich blood supply, ready for ovulation and implantation.

As the follicle grows, the cells divide and cell numbers increase, increasing the amount of oestrogen produced, which is still under negative feedback regulation. But when the follicle reaches maturity and is producing high levels of oestrogens, this triggers a switch in the hypothalamus. Scientists don't fully understand this switch, but it looks as though a special set of kisspeptin neurons in the hypothalamus may become active when

Figure 4.4 The hormonal control of the menstrual cycle in women

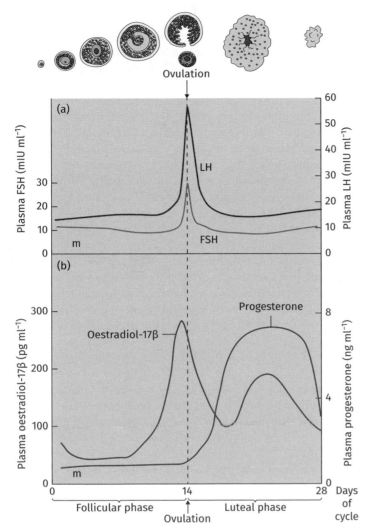

Reproduced from Pocock et al, *Human Physiology* 5/e. By permission of Oxford University Press

oestrogen levels have been high for two days. The switch means that suddenly LH production becomes *positively* regulated by oestrogens which causes a big increase in LH, called the *LH surge*. It is this surge that is the trigger for ovulation itself: the release of the mature ovum from the ovary into the Fallopian tubes, ready for fertilization.

The exact mechanism of ovulation isn't completely understood. We know that the ovum needs to get through the membrane surrounding the follicle and also needs to burst through the wall of the ovary. The LH surge appears to trigger an inflammatory cascade involving **prostaglandins** and many different local signalling molecules. The follicle contains enzymes which break down proteins. One of these enzymes, collagenase, digests collagen, which is a key component of connective tissues such as the membrane surrounding the ovary. In a coordinated way, controlled by LH, the pathway out of the ovary is cleared and the ovum is released into the Fallopian tube. Over the next few days it gradually makes its way down the Fallopian tube and into the uterus.

After releasing the ovum the ruptured follicle immediately fills with blood, but continues to develop and to produce hormones. It forms a corpus luteum, which literally means *yellow body*, in the ovary and, over a period of two weeks, produces progesterone in response to LH. The corpus luteum always has a life of two weeks unless the ovum is fertilized. The progesterone released from the corpus luteum acts on the lining of the uterus to maintain it ready for implantation if the egg is fertilized. If the egg is not fertilized, then the corpus luteum stops producing progesterone and the lining of the uterus breaks down and is shed. This is what causes menstrual bleeding.

At the time of menstruation, women are not usually fertile. There is no ovum waiting to be fertilized and the bleeding is the phase between the end of one cycle of egg release and the start of another. However, very little in biology is absolute, so although it is unlikely that a woman having sexual intercourse during her period will become pregnant, it can happen.

Dogs and rats are a bit different because they don't bleed at the end of a cycle. Anybody who owns an intact (un-neutered) female dog will know that the dog starts to produce blood spots when she is *on-heat* and therefore fertile. This is very different to menstrual bleeding. At the end of an ovulatory cycle dogs and rats do not shed the uterine lining but reabsorb it into the body.

Seasonal reproduction: sheep, caribou, Siberian hamsters, bears, and deer

Seasonal reproduction is a variant of spontaneous ovulation; it just happens that the females only spontaneously ovulate at certain times of the year. Many species of animal are seasonal breeders; this means that the males become sexually active and the females sexually receptive at a particular time of year. This ensures that mating happens at a time that allows the young to be born when the conditions (food, water, temperature) are ideal for their growth and survival. For small mammals with a short gestation period (pregnancy) and for birds, this means becoming sexually active in the early spring so that the young are born or hatched when there is abundant food: late spring and early summer.

In temperate parts of the world, like Europe and the USA, it is usually in the autumn that the larger seasonal breeders such as deer (see Figure 4.5), caribou, and bears become sexually active. Because they have a longer gestation, this means that their young are also born in the spring. In many areas one of the first signs of autumn is the sight of male deer locking antlers in a display of strength, designed to determine who is the top male. In herding animals it is very often only the top male who gets to mate with the females in his group.

So, what mechanism determines the right time of year to activate those kisspeptin neurons and stimulate both male and female animals to become sexually active? The key to understanding seasonal reproduction is a little-known endocrine gland called the pineal gland, and the hormone produced by the pineal gland, which is called melatonin. There are special melanopsin light receptors in the eye, which are different to the rods and cones

Figure 4.5 Deer such as this roe buck are a well-known example of seasonal breeders

Anthony Short

that allow us to see shapes and colours. These melanopsin receptors have a direct neural connection to the pineal gland, in the centre of the brain. In this way the pineal gland can react to changes in light and darkness in the environment. Curiously the pineal gland doesn't respond to light, but to darkness. In the hours of darkness the pineal gland releases melatonin and in the light melatonin is not produced. For wild animals then, the amount of melatonin that they produce over a single day can vary a lot between midsummer, when there is very little darkness so little melatonin production, and midwinter, when there are long nights and a lot of melatonin produced. Animals that become sexually active in autumn have a reproductive system that is programmed to respond to shortening daylength and therefore increasing levels of melatonin. The smaller animals that start to breed in the spring, such as the Siberian hamster, are programmed to respond to increasing daylength and decreasing levels of melatonin.

Melatonin implants can be used to induce early ovulation in domestic animals such as sheep. The implants, which release melatonin slowly over a period of weeks, mimic the increasing hours of darkness of the autumn and bring forward the time when sheep ovulate. By inserting an implant at the end of June, when day length is at its maximum and very little melatonin is produced naturally, sheep will ovulate in mid-August and produce lambs in January. This allows farmers to get the lambs to market weight, ready for the lucrative market around Easter.

The link between melatonin and ovulation is complex and not yet completely understood. We know that melatonin does not appear to directly regulate kisspeptin neurons, but that an intermediary is needed. There has been some recent research which has found that a part of the pituitary gland we thought was inactive, called the pars tuberalis, is in fact full of melatonin receptors. The cells of the pars tuberalis respond to melatonin by producing thyroid stimulating hormone (TSH). This TSH is regulated in a completely different way to the usual TSH that is part of the regulation of thyroid hormone, which is described in Chapters 1 and 3. Melatonin-stimulated TSH appears to just work locally in the hypothalamus, to activate the enzyme which converts thyroid hormone to its more active form. The increased local concentration of active thyroid hormone stimulates the hypothalamus and the hormone production that leads to ovulation can begin. We can probably assume that one of the targets for thyroid hormone in the hypothalamus will be the kisspeptin neurons, because they seem to be the key to all stimulation of reproductive activity, but this final link has not yet been made.

Case study 4.1

Melatonin

Melatonin is chemically different from other types of hormones: it is an indole, produced from the amino acid tryptophan (see Figure A).

Figure A Structure of melatonin

Melatonin was first discovered as a hormone that caused colour change in reptiles and amphibians. It had this effect by causing contraction of the pigment cells, called melanocytes, in reptile skin. And so it was called melatonin. It was later found that in people and other mammals, melatonin has a circadian rhythm, linked to daylight, which disappears completely in totally blind people. The link between the increase in melatonin and the start of sleep led scientists to investigate whether melatonin could be used to promote sleep and treat sleep disorders. People with dementia often have disturbed sleep but melatonin does not appear to help. There has been a lot of interest in using melatonin to help people recover from jet lag, which can cause difficulties in adjusting sleep patterns, and there is evidence that it is effective for some people. If melatonin works for jet lag, then you might think it could help shift workers to adjust their sleep patterns. Unfortunately, it doesn't seem to work for shift workers—and it isn't really clear why not.

Specific receptors for melatonin are found throughout the body, in a very wide range of organs and tissues. Some effects of melatonin are not dependent on receptors; for example, melatonin has also been found to be an antioxidant and to support immune function. It has been shown to protect DNA against damage caused by UV light and other environmental agents and so has been proposed as an anticancer treatment. As melatonin is produced in the skin as well as the pineal gland it has been suggested that it might be a natural skin protector and could be used in sun tan lotions and anti-ageing skin treatments.

Melatonin implants are also sold as hair growth promoters, for use in cattle and sheep which are being raised as show animals. Of course, it follows naturally that, if something is found to be effective in animals, it will be sold to people with promises of 'miraculous results'. This is certainly true of melatonin. If you type 'melatonin' into a search engine you will find that it is sold for a wide range of claimed benefits. It is sold to treat jet lag, sleep disturbances, hair loss in dogs, anxiety in pets, and with the general claim that it will 'improve the quality of life'. You can buy melatonin over the counter in many countries, including the USA, but it is not licensed for sale in the UK because it is considered to be an untested medicinal product. Although there is some research being carried out on the possible beneficial effects of melatonin there isn't very much work in this area. One problem is that melatonin is a natural product, so can't be patented. As a result, there is relatively little incentive for the pharmaceutical industry to invest a lot of money into researching melatonin, because it is unlikely to offer the financial return needed to justify the input of time, personnel, and funds.

❓ Pause for thought

How would you check the truthfulness of claims made online for a product like melatonin?

Induced ovulation

There is clearly an evolutionary advantage if female animals ovulate and so become fertile at a time when males are available to fertilize their eggs. Some animals have developed a system of ovulation that allows them to respond to the presence of males. In some species the females respond to the sight, sound, and smell of the males. Other species have evolved a reflex ovulation mechanism which is triggered by mating.

The male effect: goats, sheep, and deer

When female goats or sheep meet a male in prime breeding condition they respond by ovulating. This effect is most noticeable when females have been kept in isolated groups, away from males. It happens in the wild and in many domestic herding animals such as goats. The animals live in single-sex groups for most of the year and come together only to mate. This effect can override the usual seasonal pattern of ovulation in these animals and is used by people who work with livestock to coordinate ovulation in a very low-cost way (much cheaper than injections of melatonin, for example). The male effect in sheep was first reported in 1944, when we had very little understanding of exactly how it happened. The effect is most marked in species where the males compete with each other for dominance in the group and for breeding rights with the females: goats, sheep, and deer.

The male effect involves a combination of all the different senses: sight, sound, smell, touch, and taste. The most significant of these is smell. Curiously, it works across species so a farmer can coordinate ovulation in their flock of sheep by introducing a billy goat to the flock. This is very helpful when the farmer intends to use artificial insemination, rather than allow a ram to mate with the sheep. Billy goats are particularly good at having the 'male effect' because of their smell and their behaviour.

Male goats produce a **pheromone** called 4-ethyloctanal. Pheromones are volatile chemicals which can be produced by one animal and detected by another. This is different from how we usually think of the sense of smell—where we need to be aware of the smell. Pheromones do not need to have a detectable *smell* but can be recognized by a part of the olfactory apparatus. They are widely found in both invertebrates and vertebrates, and are used for chemical signalling between individuals, like a kind of group hormone.

In male mammals, the production of pheromones is often linked to testosterone levels. So, for example, the more testosterone a male goat has, the more male pheromone he produces. The pheromone is produced by glands in the skin and concentrated in the urine. The behaviour of male goats maximizes the effect of this pheromone-loaded urine. During the mating season male goats spray themselves with urine. They squat down on their hindquarters, with their penis pointing at their chest and face and urinate so that the hair on their chest and in the goat 'beard' becomes soaked with urine. Having perfumed themselves in this way they approach a female, making a snickering sound, curling their upper lip, and nudging the female in a courtship display. As a human it is quite difficult to see how any female could find this attractive—but female goats certainly do. They detect the pheromone which is one of the components of the *billy-goat* smell. The pheromone causes activation of kisspeptin neurons and generates GnRH pulses. It is a very quick effect indeed: LH increases within two to five hours of the male being introduced and the female ovulates eight to ten days later.

Reflex ovulation: cats, ferrets, and camels

Some species of animal require the act of mating to stimulate ovulation. This type of induced ovulation is known as reflex ovulation. In the ferret (*Mustela putorius furo*), the female ovary has a number of follicles which have developed to be almost ready to be released and just require that final surge of LH for ovulation to happen. The stimulus for the LH surge comes from sexual intercourse. Different species have evolved different

Figure 4.6 Camel semen contains hormone-like chemicals which stimulate ovulation in the female

RGB Ventures / SuperStock / Alamy Stock Photo

ways of initiating the LH surge. In the ferret the act of mating stimulates nerve endings in the cervix, the connection between the uterus and the vagina. The cervical nerve endings send nerve impulses to the kisspeptin neurons in the hypothalamus, stimulating release of GnRH pulses, which lead to a surge in LH. The longer that the act of mating lasts and the more frequently it is repeated, the more cervical stimulation occurs and therefore the greater the likelihood that ovulation will happen. Female ferrets usually ovulate between thirty and forty hours after mating and the ova are fertilized twelve hours after ovulation. This means that the sperm have to be fit and healthy in order to survive up to fifty-two hours after being released.

Cats also need mechanical stimulation in the act of mating so that they ovulate, but this requires a special adaptation in the male cat. When they have ejaculated, male cats withdraw their penis from the vagina. As they do this, spines along the penis scratch along the wall of the vagina, stimulating sensory nerve endings which signal to the hypothalamus. It has been suggested that this is why cats are so noisy when they are mating!

Camels (the Bactrian camel *Camelus bactrianus* and other related species), llamas, and alpacas also ovulate in response to signals during mating, but these animals don't rely on mechanical signals. The semen of the male camel contains an ovulation-inducing factor which acts like a hormone, circulating in the blood of the female (see Figure 4.6). This factor acts on the hypothalamus and stimulates kisspeptin neurons, which in turn start the hormone cascade leading to ovulation.

Scientific approach 4.1

Identification of the process of induced ovulation in camels

Camels are an economically important animal in some areas of the world where water is scarce, particularly North Africa and in parts of Russia. For successful breeding of these domesticated animals it is very useful to have information about how they ovulate. We had little understanding of how camels ovulated until the 1960s, when observation of camels led scientists to report that these animals did not appear to ovulate unless they had mated. It was assumed that camels, like cats and ferrets, needed some sort of mechanical stimulation that activated sensory nerves somewhere in the female reproductive system.

In the mid 1980s, two groups of scientists looked at ovulation in camels using the technique of artificial insemination. This involves collecting semen from males and using a small tube to insert a portion of the semen into the vagina of a female. They had four experimental groups of female camels. The first three groups had either whole semen or separated component parts of semen placed into the vagina. Group 1 received whole semen, group 2 received seminal fluid with no sperm, and group 3 received washed sperm, with no seminal fluid. The fourth group had injections of seminal fluid, without sperm, into muscle, not put in the vagina at all. Quite remarkably, camels in all the groups except group 3, ovulated.

The scientists concluded that there was something in semen, that wasn't sperm, that caused female camels to ovulate and that this factor must be transported in blood, rather than acting locally in the reproductive system. They called it ovulation inducing factor, OIF. Some scientists suggested that this OIF might be the same as GnRH. An experiment was designed to test this hypothesis. Female camels were injected with either seminal fluid or with GnRH and rates of ovulation measured. Levels of LH were also measured because these increase before ovulation can take place. They found that 83% of the camels ovulated in response to GnRH injections, but 93% ovulated in response to seminal fluid. They also found that LH levels increased within fifteen minutes of injecting seminal fluid, but only increased an hour later after the GnRH treatment. They concluded that, whatever the active ingredient in seminal fluid might be, it wasn't GnRH.

In order to try to find out what this OIF might be, the biochemists got to work. They found that the seminal fluid of camels became inactive if it was treated with trypsin (an enzyme that digests proteins) and concluded that the active constituent must be a protein. They then used gel filtration techniques that separate proteins by size and identified the size of the protein that caused ovulation. Finally, the protein was sequenced and compared to other known proteins. The amino acid sequence of OIF was found to be very similar to another protein called nerve growth factor beta (NGFβ). A biochemical technique called western blotting (see Figure A) was used to confirm this. A western blotting gel separates proteins from biological samples on the basis

Figure A This is a western blot showing that a protein found in llama seminal fluid and in the biologically active fraction of filtered seminal plasma (Fraction c2) is the same size as the positive control (NGFβ) and is recognized by an antibody that binds NGFβ. The protein has a molecular mass of 13.2 kDa.

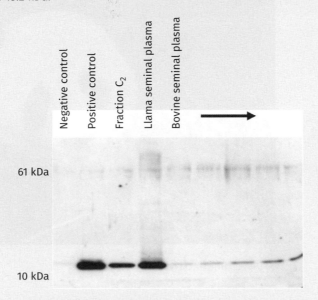

Reproduced with permission from *PNAS*. Ratto, M.H. et al. (2012) 'The nerve of ovulation-inducing factor in semen', *PNAS*, 109, 15046, fig. 5

of their molecular mass and their charge. The gel is then treated with a specific antibody which binds to the protein that we're interested in and the bound antibody can be visualized. By using a highly purified sample of the protein we can compare the purified protein to the protein in the biological sample.

Final confirmation that OIF is the same as NGFβ came from a study showing that injection of NGFβ into camels caused them to ovulate in exactly the same way as injections of seminal fluid. All this, to make camel breeding more efficient!

Evolution of ovulatory patterns

From an evolutionary perspective, you can see how the different patterns of ovulation have developed. Animals that live in mixed-sex social groups are usually spontaneous ovulators, whether they are continually cycling or seasonally active. As males are present all the time in the group they compete to win breeding rights and only the strongest gets to breed with the females. In animals that are either solitary (bears) or live in single-sex groups

Figure 4.7 People love pandas—but their unreliable reproductive strategy, amongst other things, does not bode well for their survival

ex0rzist/Shutterstock

(lions), the females ovulate in response to sexual intercourse. This allows them to breed when the opportunity is there: when a male is available.

However, there are some species with reproductive strategies which seem doomed to evolutionary failure. For example, pandas are very solitary animals (see Figure 4.7). Most solitary species have evolved to be induced ovulators, so that they can take advantage of the temporary presence of a male. However, despite being solitary, the panda is a seasonal ovulator like herd animals, which don't have restricted access to a male. For a panda to get pregnant she has to bump into a suitable male during her very short fertile period each year. With hugely reduced numbers in the wild, the chance of this happening has become very small. Even in captivity, the window of opportunity is so small and pandas are so unenthusiastic about mating unless everything is exactly right, that successful pregnancies are so rare that they still make the international news. It is perhaps not surprising that pandas are on the brink of extinction!

The bigger picture 4.1
Fertility treatments: induction of ovulation

In animals, we have seen that ovulation can be induced in several ways: introducing a male, changing day length, injecting melatonin, or by the act of mating. For animals that cycle continuously, such as humans, inducing ovulation is a little bit trickier. However, many couples have difficulty in conceiving and induction of ovulation is a useful technique for improving their chances. It is particularly useful either when there is no clear cause for their fertility problems or the cause relates to a disturbance of ovarian function, such as polycystic ovarian syndrome.

The most obvious approach is to inject the hormone LH to trigger ovulation at the normal stage of the cycle. This does work, but it is expensive and often causes problems. The best known of these is the risk of multiple ovulation leading to a particularly high frequency of multiple births. It is easier, more effective, and less risky to use tablets to change the pattern of hormone production.

The drug that is used most frequently is an oestrogen receptor antagonist called clomiphene. This drug blocks oestrogens from binding to receptors in the hypothalamus and pituitary. This stops oestrogens from having a negative feedback effect. As a result there is an increase in the pulse frequency of GnRH and an increase in the sensitivity of the pituitary gland to GnRH. This leads to an increased LH and FSH production. The result is a more effective stimulation of follicle maturation and ovulation.

❓ Pause for thought

With declining sperm counts in the developed world, about one in every ten couples have difficulty conceiving. Fertility treatment is expensive and NHS trusts in the UK are increasingly withdrawing funding from fertility clinics in order to reduce costs. They argue that fertility is not a health issue. What do you think?

4.3   Sperm production

All male animals need to produce sperm to father offspring and pass their genes to the next generation. Sperm are produced in the testes from the time of puberty right up until death. Men are not born with a fixed number of sperm but make new sperm all the time. Special cells in the testes, called spermatogonia, produce immature sperm which mature as they pass through the labyrinthine tubes of the testes. Both processes, of sperm production and maturation, need hormones; FSH is a key hormone (see Figure 4.8) and high levels of androgens, particularly testosterone, are also needed. Testosterone here is not acting as a hormone because it doesn't need to get into the blood to have its effect. It is acting as a local regulator. You might expect that testosterone, being a small and very lipid-soluble hormone, would leak out of the testes, moving down its concentration gradient, into the blood, but one of the effects of FSH is to produce a special protein which binds to the testosterone and keeps it in the testes.

We saw in Chapter 3 that an important part of the development of a baby boy is the movement of the testes out of the abdomen and into the scrotum.

Figure 4.8 Hormones controlling sperm production. The hormones from the pituitary gland, LH and FSH, are both needed for sperm production.

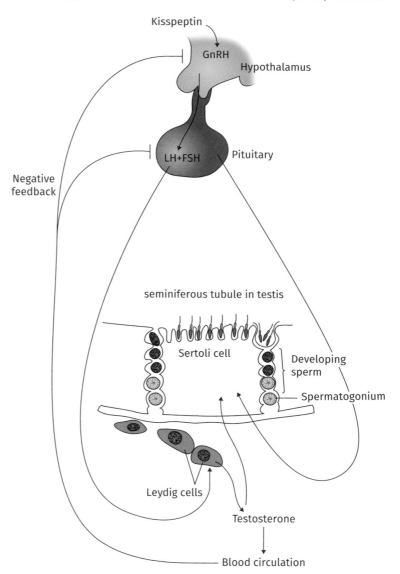

The scrotum is a pouch of skin that is outside the main part of the body. It is quite an exposed place for the testes to be found, outside the protection that other important organs have. This makes the testes quite vulnerable to damage (see Figures 4.9 and 4.10). There has to be an evolutionary drive to explain why they are found in such a vulnerable place. The answer to this conundrum appears to be temperature. While core body temperature is 37°C, the temperature inside the scrotum is typically two degrees lower, at 35°C. The

Figure 4.9 In the scrotum, outside the abdomen, the testes are vulnerable to damage. When you watch a football match and see players lining up to face a free kick, many of them will have their hands placed in front of them to protect their scrotum.

Hind Sight Media / Alamy Stock Photo

temperature is partly regulated by a group of muscles which tighten or relax the scrotum, depending on the outside temperature. In the cold the muscles tighten, pulling the scrotum closer to the abdomen. The lower temperature of the scrotum appears to be necessary for normal sperm production. When it was first reported that sperm counts of men in the developed world were declining, the decrease was blamed (without any evidence) on men wearing tight underpants, and boxer shorts suddenly became fashionable!

During their life men produce billions of sperm. Each ejaculate contains between forty million and 1.2 billion sperm in a volume between 0.5 and 10 cm^3. This is a relatively modest volume of ejaculate; boars typically produce up to 300 cm^3! In fertile men the concentration of sperm is usually above fifteen million per cm^3. Levels below this are likely to cause fertility problems. That doesn't mean that a man with a low sperm count will not be able to father a child, but that it will be more difficult, even though it does only take one sperm to fertilize an ovum.

Given that it only requires one sperm to fertilize an ovum, why do almost all males, across the animal kingdom, produce such a huge excess of sperm?

Numbers of sperm is not all that matters in male fertility; the percentage of abnormal sperm is also important and so is the swimming ability of the

Figure 4.10 As you can see from this picture of a male dolphin, some male animals do not have a scrotum but keep their testes inside the abdomen. What adaptations must have evolved to make sperm production possible?

Neirfy/Shutterstock

sperm. When sperm count is checked in a fertility clinic these two additional factors are used to measure *sperm quality*. Sperm have a lot of swimming to do if they are to get from the vagina up into the uterus and fertilize an ovum, beating all the other sperm in the race to be the first.

4.4 Contraception

People have always sought to control their environment and to improve their own living conditions. Seeking to control our own fertility is an aim that started with the very early civilizations. The simplest forms of contraception are barrier methods, such as the condom, which act as a physical barrier between sperm and ovum. The use of these barrier methods goes back thousands of years, to when pieces of animal gut were used as rudimentary condoms. However, it may surprise you to learn that oral contraceptives were in use thousands of years ago as well. The ancient Greeks

Figure 4.11 Silphium, the herb shown on this Roman coin became extinct in the third or fourth century AD because it was such an effective oral contraceptive

The History Collection / Alamy Stock Photo

used plant-based contraceptives; both pomegranate seeds and a herb called pennyroyal (*Mentha pulegium*) were eaten to prevent conception. Another herb, called silphium, was a highly effective contraceptive in both Greek and Roman times, leading to such great demand for the plant that it became extinct in the third or fourth century AD (see Figure 4.11). A related plant, asafoetida, although less effective, is still used in modern times. These plants were not taken regularly in the way that the oral contraceptive pill is used today, and they don't appear to prevent ovulation. Instead, it seems that they were an early form of *morning-after* pill, used after sex to prevent fertilization or implantation of the conceptus.

Until the development of modern hormonal contraceptives in the mid-twentieth century the only alternatives to barrier methods were withdrawal (sometimes called coitus interruptus), where the man withdraws his penis from the vagina immediately before ejaculation, and the *rhythm method* of only having intercourse during the time of the menstrual cycle when pregnancy is less likely. Withdrawal is not an effective way to avoid pregnancy because some sperm are released during intercourse, before ejaculation. The rhythm method of contraception involves predicting the day of ovulation and avoiding sex around this day. There are two ways of predicting ovulation. The first is to use a calendar and count from the date of the start of the last menstrual period. As we saw earlier in this chapter, this is the time interval that varies during the menstrual cycle. For some women the interval between menstruation and ovulation appears absolutely fixed, and they have an unvarying cycle length of twenty-eight or thirty days. Many women do not have a fixed cycle length and the menstrual cycle varies between twenty-six and thirty-five days almost on a month-by-month basis. This makes a calendar-based prediction of ovulation very unreliable as a method of avoiding unwanted pregnancy. The second method is to use a calendar combined with a thermometer, based on the fact that ovulation is associated with a small but measurable rise in body temperature. Of course, this means that you can only determine when ovulation has taken place, rather than predict when it is going to happen. As sperm can survive for several days in a woman's reproductive tract this method is more useful for planning when to have sex in order to achieve conception, rather than preventing it. Of course, there are now apps that help women track their ovulatory cycle. Do you think that these are likely to be any more reliable than marking a date on a calendar?

In the mid-twentieth century the first hormonal contraceptive was produced commercially and made widely available to women in the developed world. Apart from a reduction in the dose of steroids, this product was exactly the same as the contraceptive pill used today. It consists of a tablet, taken daily, containing a mixture of two steroids which act like an oestrogen and progesterone. This is a combined oral contraceptive pill. The oestrogen acts on the pituitary gland to stop release of LH and FSH and so prevents ovulation. It also causes the mucus produced by the cervix to become thick and sticky, forming a physical barrier to sperm. The progesterone is included because it has two key actions: it makes the oestrogen more effective and it prevents unwanted effects of oestrogens, including increased cancer risk. There is a form of oral contraceptive called the progesterone-only pill, which is also effective in blocking ovulation, but has to be taken at exactly the same time every day, whereas the combined pill allows more flexibility.

The steroids used in the contraceptive pill are synthetic, but made to be similar to natural oestrogens and progesterone. They work in a one-a-day form because of the amazing way the body handles steroids. When we swallow a contraceptive pill it dissolves in our stomach and small intestine and the steroids are quickly absorbed into the bloodstream. The liver metabolizes these steroids as the blood from the intestine passes through it, and some is excreted in the urine. Much of the metabolized steroid, however, is sent from the liver to the gall bladder and ends up back in the gut, where the process of metabolism is reversed and the steroids are once again absorbed into the bloodstream. This keeps happening over and over again and the cyclical process keeps the steroids recirculating in the body throughout the day, keeping blood concentrations of the steroids fairly constant and making the pill more effective. The oral contraceptive pill has a very high success rate, but only if it is used correctly, and that means remembering to take it every day! Of course, if we have a stomach upset and have diarrhoea, then the gut contents are lost before the reverse metabolism and reabsorption can take place, so the cycle is broken, the blood concentrations of steroid can fall, and the pill may stop being effective.

Steroids are also used in other forms of contraceptive; implants made of synthetic progestogens are most commonly used. These are in the form of small rods which are inserted under the skin and which release the hormone very slowly over a period of several months. These implants work in the same way as the pill, stopping ovulation and thickening the cervical mucus. They are even more reliable than the oral contraceptive pill because you can't forget to take one!

There are also hormonal implants used in the modern coil, or intra-uterine system. The old type of coil used copper, which acted as a local irritant and stopped a fertilized ovum from implanting in the uterine wall and developing. The more recent versions are made of plastic, with a system for slowly releasing progesterone over time. The progesterone is released locally, in the uterus, and so causes fewer side effects than the oral contraceptive or long-term implant. Again, it works by causing thickening of the mucus in the cervix, preventing the sperm from reaching an ovum. The progesterone also causes the lining of the uterus to become thinner, so a fertilized egg can't implant. One additional benefit of this effect is that menstrual periods become much lighter, which is a really positive thing for most women.

Case study 4.2
Contraceptive pill for men

You will have noticed that all the hormonal contraceptive methods we have described are targeted at a woman's reproductive system. Women either take pills, have implants, or use an intra-uterine system. So why isn't there a hormonal contraceptive option for men?

A good starting point is to look at what a contraceptive needs to do:

- It has to be effective at preventing pregnancy.
- It must be safe, with only very minor unwanted side effects.
- It must be reversible because most people want to control their fertility, not lose it completely.
- It shouldn't interfere with the act of lovemaking.
- Ideally, it should be low cost and easy to use.

Many different approaches have been tried over the past thirty years.

You might think that it would be easiest to copy the approach used for the female contraceptive pill and put androgens, the male sex hormones, into a daily pill. There are two problems with this: one is that androgens are very quickly deactivated by the liver and not very well reactivated in the gut. So the hormone cycle that works so well in women doesn't work in men. The other problem is that the liver becomes damaged by these steroids when taken as a pill.

An alternative approach was to try a GnRH mimic. This might seem a bit il-logical, because GnRH *stimulates* reproductive hormones. But it only has this effect when released in pulses with a ninety-minute interval between pulses. When it is injected, so that there is a constant level in the blood, it works quite well at blocking LH and FSH release. This stops sperm production in most men, but it also blocks testosterone production. A lack of testosterone is a significant problem. It causes tiredness, muscle weakness, loss of sex drive, and can mean that men don't get an erection. To try to get round this problem, researchers have tried combining the GnRH mimic with an andro-gen. It works, but the need for frequent injections means that it's not a very practical option for most men.

One of the more successful products developed has been a combination of progesterone, as in the female contraceptive pill, with testosterone. While the progesterone can be taken orally, the testosterone is given as a patch. This combination of steroids stops sperm production in nearly all men, has very few side effects, and is quickly reversible. So why isn't it available to use? The answer is simple; pharmaceutical companies are reluctant to invest money in developing new ways for men to control their fertility and view the restoration of potency (e.g. Viagra) as a much bigger market. Until that changes, men who wish to control their own fertility are stuck with a poor range of options: they can use a condom or have a vasectomy. Neither of these tick all five of the requirements for a contraceptive.

❓ Pause for thought

Which part of the male reproductive system would you target if you were de-signing a male contraceptive?

Globally, it is easier to reach women with contraceptive advice than men. Suggest why this might be and how it could be addressed.

4.5   Pregnancy and birth

Nearly all mammals have a time of pregnancy in their reproductive cycle. The only exceptions are the very few species of mammal which lay eggs (the duck-billed platypus and echidna). Pregnancy, or gestation, is the name given to the period of time when the fertilized ovum develops into a baby animal inside the mother's body. The end of pregnancy, the time when the young animal leaves its mother's body, is called birth or in scientific terms, parturition. The length of pregnancy (also called the gestation period) var-ies hugely between different species, partly according to the size of the ani-mal and partly depending on how well developed the baby animal needs to be before it is born. So a mouse has a gestation period of twenty-one days and the baby mice are born naked, blind, and unable to move around very much. Kittens are born after sixty-four days of gestation and are also blind and quite helpless. Humans have a gestation period of around 270 days and babies, although not blind, are not able to walk or feed themselves for several months. A baby lamb, on the other hand, is born after 147 days, and is able to very quickly stand up and follow its mother. The longest gestation period is the African elephant, with a 630 day pregnancy; not far off two years. The sperm whale also has a long pregnancy, lasting 535 days, or nearly eighteen months. The young of both these animals are able to move and follow their mother very quickly after birth.

Mammals are the only group of animals that go through pregnancy. As an adaptive mechanism it has huge advantages: the young are protected and kept in an absolutely perfect environment for them to develop. The organ that allows pregnancy to happen is the placenta.

When a fertilized ovum embeds itself in the uterine wall this is the start of pregnancy. The developing embryo has to establish a good blood supply to make sure it gets enough oxygen and nutrients. The placenta is the organ that supplies the embryo with blood. It forms the interface between the mother and the developing foetus and is a truly remarkable organ. The placenta is developed from the foetus so it is *foreign* to the mother, but her immune system doesn't react against it. We are still not entirely sure why. It is also an endocrine organ. As well as making its own set of hormones, the placenta modifies the mother's hormones and ensures that the hormonal environment protects the foetus.

The first hormone that the placenta makes is called hCG or human chorionic (pronounced core-e-onnick) gonadotropin. This is the hormone that is detected in the pregnancy test (see Chapter 2). The hormone hGC is not made anywhere else in the body so is a good *marker* of pregnancy. It takes over from LH to keep the corpus luteum active and making progesterone through the first part of pregnancy. Later in pregnancy the placenta takes over progesterone production itself and continues to produce progesterone all the way through to the end of pregnancy. Progesterone is important because it keep the muscles of the uterus relaxed until right at the end of pregnancy. The placenta also inactivates the stress hormone cortisol, produced by the mother, so that the foetus is protected from potentially harmful exposure. As the placenta grows it produces increasing amounts of a hormone called placental lactogen, which prepares the breasts for milk production.

At the end of pregnancy the hormone production by the placenta changes significantly. The placenta produces much less progesterone, allowing the uterine muscle to respond to signals, causing muscle contraction. It starts producing prostaglandins, which prepare the cervix for opening to allow the baby through. The actual signal to start the process of parturition (giving birth, also called *labour*) depends on the type of animal. In sheep the adrenal glands of the lambs send signals to the mother to start the birth process. In humans it is another hormone from the placenta—CRH, which usually comes from the hypothalamus and starts the stress response (see Chapter 2). CRH is made by the placenta in the final weeks of pregnancy and appears to be the key hormone for starting parturition.

At the start of parturition, as progesterone levels drop and the uterus starts to contract, levels of oxytocin increase. Oxytocin is a small peptide hormone from the pituitary gland. It makes certain types of smooth muscle contract and is a key hormone in parturition. Although the uterus is contracting before oxytocin is released, the presence of oxytocin makes the muscles of the uterus contract in a coordinated way, pushing the baby through the cervix and out of the mother's body. On maternity units in hospitals, injections of oxytocin can be used to induce labour if a baby is overdue. It is combined with prostaglandins, given as pessaries placed next to the cervix. The prostaglandins help the cervix to soften and relax, while the oxytocin increases the strength and frequency of uterine contractions.

The placenta continues to supply oxygenated blood to the baby until it takes its first breath of air. Then the placenta's job is done and it comes away from the uterus and is delivered after the baby, often called the *afterbirth*.

Hormones and milk production

Hormones continue to have an important role after the baby, or kitten, or puppy is born. All mammals feed their babies with milk produced in their own bodies. The glands that produce milk are properly called mammary glands but are usually called breasts in humans (see Figure 4.12).

Figure 4.12 Breastfeeding has huge health advantages for both mother and baby

Blend Images / Alamy Stock Photo

There are two hormones that work together during lactation (milk production). They are both from the pituitary gland and produced in response to a baby, or kitten or lamb, sucking its mother's nipple. Sensory nerves from the nipple send signals to the hypothalamus, which then causes hormone release from the pituitary gland. The first hormone, prolactin, works to increase milk production. The second hormone is oxytocin. We have already seen how this hormone causes smooth muscle contraction in the uterus, and it also has this effect in the breast, causing the contraction of smooth muscle around the milk ducts. This has the effect of pushing the milk towards the nipple and is called the *milk let-down reflex*. Sometimes breastfeeding mothers find that just thinking about their baby releases oxytocin and causes milk to leak from their breasts. It is interesting that oxytocin can be produced without the sensory nerve stimulation caused by a baby actually suckling. It is also curious that men produce oxytocin; why would a man produce a hormone that causes contraction of muscles in the uterus and milk ducts when he hasn't got either of these tissues? It turns out that there is a lot more to oxytocin than its role in parturition and lactation. We will look at this hormone more closely in Chapter 5.

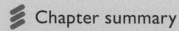 Chapter summary

- Both male and female reproductive tissues produce steroid hormones as well as gametes.
- Growth during childhood is under endocrine control, and involves growth hormone, thyroid hormone, adrenal steroids, and insulin.
- Puberty is a process that is initiated by the development of kisspeptin neurons, which signal the onset of menarche in girls at a body weight of around 47 kg; but the exact mechanism is uncertain in boys.
- Ovulation in different animals may be on a continuous cycle, on a seasonal cycle, or induced by contact with a male.
- A woman has only a limited number of ova that mature during her reproductive life, whereas men continuously produce new sperm, from puberty until they die.
- Hormones are essential for pregnancy to occur, continue normally, and result in delivery of a healthy baby. The principal source of hormones in pregnancy is neither the mother nor the foetus, but the placenta.

Further reading

https://www.fpa.org.uk/sites/default/files/your-contraceptive-choices-chart.pdf
This NHS website has useful information about hormonal and non-hormonal forms of contraception.

https://www.ncbi.nlm.nih.gov/pubmed/26732151
This link takes you to a published review article about the options for developing a hormone-based male contraceptive.

https://www.hfea.gov.uk/treatments/
The Human Fertilisation and Embryology Authority website has some very useful information about fertility treatments.

Hinson, Raven, and Chew: *The Endocrine System* (Churchill Livingstone)
This textbook is aimed at medical students but has a lot of information about human reproductive biology.

 ## Discussion questions

4.1 How much do you think that the introduction of reliable hormonal contraception in the 1960s contributed to changes in attitudes to sex?

4.2 What mechanisms might be responsible for the initiation of puberty in boys, given that it does not appear to be a direct association with weight as it is in girls? How could you investigate this?

4.3 Think about some of the animals not mentioned in this chapter and try to work out their ovulation pattern, based on their social behaviour (e.g. solitary or herd).

5 HORMONES AND BEHAVIOUR

The very earliest experiments in endocrinology were Berthold's work with chickens, which we met in Chapter 1. They involved watching animal behaviour. Chickens with testes behaved differently from chickens without testes. People have actually known for thousands of years that hormones can alter behaviour. Roman gladiators ate bulls' testes to make them better fighters; as well as increasing strength, the testes were thought to give courage. Of course, in Roman times they hadn't heard of testosterone, but they recognized the principle that some element within the testes assisted gladiators. Despite this early knowledge, the behavioural effects of many hormones are still being uncovered. We are only just beginning to understand that hormones have important roles in complex social behaviours that were previously thought to be the research area of psychologists, not biologists. In this chapter we will look at some remarkable stories of hormones and behaviour. We will take a fresh look at the two hormones released by the posterior pituitary gland: oxytocin and vasopressin. We will look at the role of hormones in controlling our food intake and, finally, examine the effects of testosterone on the brain.

5.1   Oxytocin and vasopressin

Oxytocin and vasopressin are both short peptide hormones produced in the hypothalamus and released from nerve terminals in the posterior part of the pituitary gland. They have evolved as a result of a gene duplication in an ancient vertebrate species. This ancestral gene has given rise to a family of related peptides in the different types of vertebrates, but ultimately to oxytocin and vasopressin in mammals. Curiously, the genes for oxytocin and vasopressin are both on the same chromosome, but oriented in opposite directions. Gene mutations during vertebrate evolution have caused changes in the amino acid sequence from the original in some positions, but oxytocin and vasopressin are recognizably related peptides. Their receptors are also very similar in structure and so there is some overlap between their actions. This makes studying their effects on behaviour quite complicated.

As we saw in Chapter 4, oxytocin has a role in human childbirth and breastfeeding. The hormone acts on smooth muscle in the uterus to cause contractions, which are the key to delivery of a baby. After the birth, when the baby suckles at the breast, oxytocin is the hormone that causes the smooth muscle in the milk ducts to contract and allows the milk to flow properly. Until about twenty years ago, as far as we knew, oxytocin was only important in childbirth and in breastfeeding. There were a lot of questions that puzzled scientists about oxytocin; why did men produce oxytocin? And why did women produce oxytocin when they weren't giving birth or breastfeeding? We now know that oxytocin is much more than a hormone of childbirth. It has profound effects on the brains of both men and women, not to mention birds, mice, and many other animals as well.

We are already familiar with the actions of vasopressin as a hormone which works to conserve water in the body by enabling the kidney to produce concentrated urine, and we saw in Chapter 2 how a defect in the receptor for this hormone results in a disorder called diabetes insipidus. We now know that, like oxytocin, vasopressin has a range of actions in the brain as well as its physiological effects on the kidney.

Case study 5.1

Vasopressin: more than just the antidiuretic hormone

The fish version of vasopressin is called **vasotocin**. Like vasopressin in mammals, vasotocin plays an important role in water balance in fish. It appears to be especially important in species of fish which are semi-amphibious, enabling them to adapt physiologically to spending part of each day out of the water. The mudskipper (*Periophthalmus modestus*) is a good example of a semi-amphibious fish. This fish lives in the brackish, semi-saline water of river mouths in Japan, and at low tide feeds on the exposed mudflats, out of the water. Mudskippers are also very aggressive fish. Surprisingly for a fish, males establish a social hierarchy through displays of aggression on land, in a similar way to other vertebrates and, of course, it is only the dominant males who get to mate with the females.

The aggressive displays of male mudskippers start with raising their dorsal fins like sails on their backs. This increases their apparent size. They then face their opponent with their mouth wide open, again appearing to increase their size. The next step is to move towards their opponent with fin raised and mouth open; as the display escalates they move more quickly and try to force the opponent to give way. More aggressive males will actively chase a fleeing opponent and may even attack and bite them. Each encounter between two males results in one emerging as the dominant individual, with the other being subordinate (see Figure A).

Figure A Mudskipper males fight to establish dominance. Injections of vasotocin make them more aggressive.

aDam Wildlife/Shutterstock.com

Scientists in Japan carried out a series of experiments to investigate the role of vasotocin (VT) in the aggressive displays of these feisty fish. They took a pair of male mudskippers of the same size and injected a small amount of VT into the brain of one, and a small amount of a control fluid into the brain of the other. They placed the two fish in a tank set up to resemble the mudflats and watched what happened. The scientists counted the numbers of aggressive acts of each fish: fin raising, mouth opening, chasing, and biting. They found that the males who had received the VT injection were far more aggressive than the control males, who hadn't received VT. These VT-treated males showed a threefold increase in fin raising and in chasing, and a sixfold increase in biting and attacking. Scientists compared the effects of two different doses of VT and found that the high-dose treatment resulted in more aggressive acts than the low-dose treatment. Where increasing doses of a compound have increasing effects like this, the response is called **dose dependent,** and this is strong evidence that there is a causal link. In this case, it seems clear that VT has a role in stimulating aggression in mudskippers (see Figure B).

Figure B This graph demonstrates the effects of low-dose and high-dose VT treatment compared with controls on levels of aggression in mudskippers. Each bar shows the mean +/- standard error of observations on six fish.

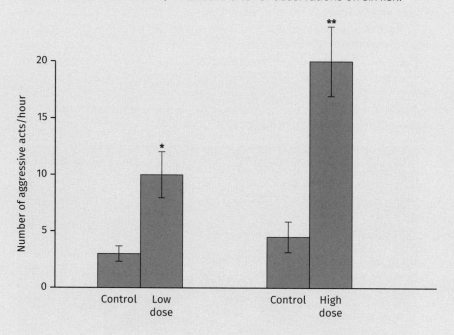

Oxytocin: the relationship hormone

Oxytocin is released when we have affectionate physical contact with another person. In babies and their mothers, we know that oxytocin levels increase in the mother when the baby suckles at the breast. This enhances what we call the *maternal bond*, which is the sense of love and protectiveness that a woman feels for her baby. But even without suckling, just a cuddle causes oxytocin levels to increase in both mother and baby. The effect is even greater when they have skin-to-skin contact. In recent years, in maternity hospitals, midwives have recognized the importance of a baby being delivered directly onto its mother's stomach, rather than being whisked away, cleaned up, and wrapped in a blanket before being handed to the mother. This simple difference has an effect on how well mothers and babies *bond*. Improved bonding has long-lasting effects on the wellbeing of both mother and baby. So during breastfeeding, oxytocin seems to be having effects both on the brain and on smooth muscle contraction.

There is one other situation where oxytocin seems to have this dual effect on both the brain and the smooth muscles. In couples who have a sexual relationship, oxytocin levels increase not just in response to cuddling, but also as a result of sexual climax or orgasm. There is an interesting sex difference: women have a much greater release of oxytocin following orgasm than men. One theory is that the oxytocin may be acting on the smooth muscle of the vagina and uterus, helping to draw the ejaculate and sperm towards the ovum and so assisting in conception. In both women and men, the oxytocin released appears to increase their affection for each other. In behavioural terms this is called attachment, and this appears to be an important factor in maintaining relationships.

Biology demonstrates that cuddles count

We have seen how, in babies, oxytocin release can be increased by physical contact between mother and baby. The effect is seen in people of all ages, and does not just involve mothers and their children. In older children and adults, a hug from a friend or family member causes an increase in oxytocin levels in the blood. Even pets have an effect on our oxytocin levels and we do on theirs; stroking your dog or cat can make oxytocin levels rise in both the pet and the human. You can see why oxytocin is sometimes called the *cuddle hormone*. In a way this is misleading, because it seems that the physical contact isn't always necessary. One study found that teenagers, studying away from home, experience a rise in oxytocin when they talk to their mother on the phone. You could describe oxytocin as the glue that holds human societies together; any positive interaction between people increases oxytocin levels and increased oxytocin levels improve our sense of belonging and our attachment to other people. The flip side of this is that people who have experienced abuse during their childhood have an impaired oxytocin response and produce much less oxytocin in response to a hug or other positive interaction.

Clearly, oxytocin released in this way is having quite a different effect to the actions on smooth muscle that we are familiar with, in childbirth and in milk let down. The neurons in the brain that make oxytocin (and vasopressin) are found in two nuclei, or clusters of brain cells, in the hypothalamus. These nuclei are the supraoptic nucleus (SON) and the paraventricular nucleus (PVN). Cells in both the SON and PVN have axons that extend into the posterior pituitary and it is from here that the hormones are released into blood. But there are also projections from both nuclei into other parts of the brain, including the parts which affect behaviour. Oxytocin has both local effects on the brain and hormonal effects on the body when it travels through the blood. It appears that the effects are linked because, as we have seen, blood concentrations of oxytocin change in response to behavioural stimuli.

Scientific approach 5.1

The reading the mind in the eyes test (RMET) and other tools for measuring behavioural effects of hormones

In the 1990s, a group of scientists at the University of Cambridge, led by Dr Simon Baron-Cohen, devised a simple test of a person's social intelligence. Social intelligence is sometimes called empathy or emotional intelligence: the ability to 'read' another person's emotional state. The test is called the *Reading the Mind in the Eyes Test* (*RMET*) and allows the scientist to evaluate how well somebody can understand another person's emotional state if all they can see is the person's eyes (see Figure A). The test involves showing somebody a series of pictures of people's faces showing only the eyes and asking them to guess the emotion that is being expressed. This study formed part of their work on **autistic spectrum disorder** and has led to further research looking at the effect of several factors, including hormones, on the results of the test. Autistic spectrum disorder is associated with impaired social functioning and communication and so people with autistic spectrum disorder find it more difficult to guess the emotions being shown in the pairs of eyes. There are other factors that affect performance in the RMET as well, including sex. Women generally score more highly than men in this test and this has led to research to try to explain this finding.

Figure A It is really easy to tell that this man is happy when you can see his whole face. How easy is it to work out his emotion when you can only see his eyes? This is the basis of the Reading the Mind in the Eyes Test.

SFIO CRACHO/Shutterstock.com

Scientists have looked at the effects of oxytocin on how well people perform in the RMET. Oxytocin is very easy to administer to people; as it is such a small peptide it is readily absorbed from mucous membranes, so it can be given as a nasal spray. Oxytocin has very little effect on most people; it does not appear to close the gap between the performance of men and women. However, when given to people diagnosed with Asperger syndrome, one of the autistic spectrum disorders, it significantly improves their score in the RMET, suggesting that it can improve social intelligence in people with impaired social skills. There have been several other studies looking at the effect of oxytocin on children, adolescents, and adults with autistic spectrum disorder. In general, these studies have been quite short term and have involved small numbers of people, but the results have been promising. It seems that oxytocin may improve communication and social functioning, including facial recognition, in some people with autistic spectrum disorder. This is an important finding because people with autistic spectrum disorder are often impaired by their poor social skills, finding it hard to interact with other people socially, at work, or in their personal lives. There is still a lot of research that needs to be carried out to answer some of the questions. Does oxytocin continue to work if it is used for a long period of time? Do the benefits continue after treatment is stopped? What determines which people respond to oxytocin and which don't?

In the same way that crowd funding has revolutionized the way that new businesses start up, the internet has also allowed interesting new ways of carrying out research on a mega scale. These studies usually involve very large numbers of participants who complete online questionnaires or tests. For example, Harvard University is doing a big study looking at how people score on the RMET. If you would like to take part in a worldwide research study and find out how good you are at reading emotions, you can look at the further reading at the end of this chapter for a link which invites you to take part in this research.

There are other behavioural tools used by scientists to evaluate the effects of hormones. For example, risk-taking can be measured using a game based on gambling and trust, using the *Prisoner's Dilemma*. This model requires two people: they are asked to imagine that they are both prisoners. They are put in separate rooms and each offered the following choices. If they both keep silent they will have one year in prison. If they both betray the other they will both get two years in prison. However, if prisoner A betrays prisoner B, while prisoner B keeps silent, then prisoner A is released and prisoner B has three years in prison. The decision-making in this model gives a measure of how much trust each person has. The effects of oxytocin administration on trust have been measured using this model. The first studies that were carried out, using small numbers of people, reported a positive effect of oxytocin, but these experiments have not been very reproducible. Larger studies have not found significant effects, although overall there is a lot of evidence that oxytocin administration can increase trust in a variety of experimental situations. Perhaps, as we have seen with the RMET previously described, there may be some people who respond to oxytocin in a particular situation, while others do not. It may be the case, as some researchers have suggested, that, in a small number of people, there is a clinical state of *hypo-oxytocinism* which results in antisocial and aggressive behaviour, and which can be effectively treated with oxytocin. It is clear that there is still a lot that we don't know about the role of oxytocin in human social interactions.

❓ Pause for thought

Many food shops use the smell of coffee or baking bread to entice shoppers to spend more money. We accept this without complaint. Would you be concerned if a bank, for example, sprayed oxytocin in the air, in an attempt to make customers trust them more?

Oxytocin in other species

The behavioural effects of oxytocin have also been studied in other species with some really interesting results. The species most studied are those that form social groups. Studies on birds and small mammals found a new and unexpected effect of oxytocin on behaviour: it has a role in sociability. Zebra finches (*Taeniopygia guttata*) are sociable birds, gathering in flocks to feed and sleep. Treating female finches with an oxytocin antagonist, or blocking oxytocin production, makes them less sociable. In male zebra finches the same oxytocin antagonists make the birds overeat. This is a very interesting observation that we shall return to when we look at hormones and eating behaviour (see Figure 5.1).

Voles (*Microtus sp.*) are also often used to study the role of oxytocin in behaviour. The females of some species of vole are **monogamous**, which

Figure 5.1 Zebra finches and voles have both helped us to understand more about the effect of oxytocin on behaviour

Simona Bottone/Shutterstock.com   Jon_Clark/Shutterstock.com

means they form a relationship with one mate and stay with that mate for life. Other vole species are non-monogamous and mate with several different partners. Scientists have found that treating the monogamous voles with an oxytocin blocker makes them behave like the non-monogamous voles. The brains of the two types of vole have been studied and it has been found that the brains of the monogamous voles have more receptors for oxytocin than those of non-monogamous voles. Oxytocin seems to be critical for pair bonding in these animals. Scientists went on to look at the effects of separating pairs of monogamous voles. When separated for short times, oxytocin levels decreased, but then increased when the pair were reunited. This seems to be the hormonal equivalent of missing one another. Long-term separation caused a complete crash in oxytocin signalling, but giving the voles oxytocin improved their coping mechanisms and made it more likely that the vole would find another partner. We know that in the brain, oxytocin signalling has direct links to the part of the brain concerned with *reward* and pleasure. Scientists have suggested that this reaction in voles may be an animal model of bereavement. They are exploring the question of whether altering oxytocin levels in people may help them cope better with the death of a partner.

In addition to its effects on coping mechanisms, pair bonding, and reward, it is thought that oxytocin may have a link to the immune system. It is suggested that people with a higher oxytocin level may have a healthier immune system. So it is possible that oxytocin may link bereavement, loneliness, coping behaviour, and the immune system.

5.2   Hungry hormones: how hormones control our eating behaviour

Eating is a complex behaviour. We eat as part of a social situation, we eat because food looks or smells nice, and sometimes we eat when we're feeling miserable: comfort eating. However, at the most basic physiological level we, and other animals, eat when our body signals to us that we're hungry. We're all used to the idea that we get hungry when it's time to eat and stop feeling hungry when we have had a meal. It's not just whether or not there is food in our stomach that makes us feel hungry, we can have an empty stomach and not feel hungry. It's not even a question of blood glucose levels. There are, of course, hormones involved. These hormones signal to the brain to tell us when we should eat, how much we should eat, and when to stop eating (see Figure 5.2). Hormones even control the types of food that we eat.

The hormones you can see in Figure 5.2 all act on the brain, sending signals to the hypothalamus. This is the area of the brain that regulates hunger, thirst, and temperature, and which also coordinates the release of several hormones. Some hormones act directly on the hypothalamus, but others affect the vagus nerve. This is part of the autonomic nervous system involved in parasympathetic (unconscious) control of the heart and digestive system. The vagus nerve also has a sensory input to the hypothalamus.

Figure 5.2 Hormones act on the brain as part of a complex system to regulate hunger

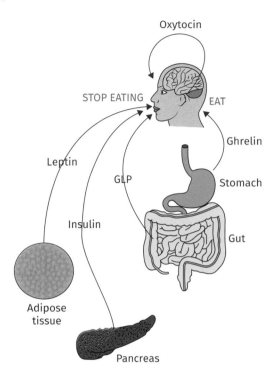

The hormones that control how hungry we feel don't come from the *classical* endocrine glands, with the exception of insulin. Mostly these hormones come from the stomach and the gut. One of the more interesting hormones in the control of hunger is called **ghrelin** (pronounced gree-lin). This peptide hormone is produced by cells in the stomach when the stomach is empty. You have stretch receptors which are activated when your stomach is full and they stop the release of ghrelin. Ghrelin acts on the brain to stimulate hunger and cause increased food intake, and then blood levels of ghrelin decrease during and after a meal. There is also a circadian rhythm of ghrelin release which shows a gradual increase during the night-time when we are asleep, although disturbed sleep and shift work both disrupt the normal pattern of ghrelin release. This may contribute to obesity as the increased ghrelin drives hunger when we are sleep deprived. Scientists have shown that abnormal ghrelin production is not a major cause of obesity because people who are very overweight generally have lower levels of ghrelin than thinner people. There is research into whether a ghrelin blocker might be an effective weight loss treatment but, although it appears to have some effect in rats, no studies have yet been carried out on people.

As well as the hormones that make us feel hungry, there are others that tell our brain when to stop eating. Insulin is one of these hormones, another is leptin, and our old friend oxytocin is on this list as well. We'll start by looking at leptin. Leptin was discovered as a result of a gene mutation in a laboratory mouse. This mouse came from a family of lean mice, but ate voraciously and became obese. Its offspring were also obese. The mutated gene was called the *ob* (short for obesity) gene and the strain of mouse called *ob/ob*. In the 1990s it was discovered that this gene which caused obesity coded a protein that circulated in blood. The protein was named leptin from the Greek for *thin*. The *ob/ob* mice produced a mutated form of leptin that did not activate the leptin receptors in the brain; they were, in effect, leptin deficient, and this resulted in obesity.

Leptin is a polypeptide hormone which is produced by our fat cells (see Chapter 1, Table 1.1). The hormones produced by fat cells (properly called adipocytes) are called adipokines, and leptin is just one of a whole range of signalling molecules made by these cells. The more adipocytes we have, the more leptin we produce. The fat cells produce leptin, which circulates in blood and binds to receptors in the arcuate nucleus of the hypothalamus, inhibiting hunger signals. When leptin is injected into *ob/ob* mice it stops them from overeating. It had exactly the same effect in *wild type* mice (mice that don't have the *ob/ob* mutation). Leptin is also produced by human adipocytes, and it has been shown that it acts to suppress hunger in humans as well. Physiologically, leptin acts to regulate our fat stores around a set point. When our body fat increases and our leptin levels rise above the set point, leptin acts to suppress our appetite and helps us to maintain a constant level of body fat, and so a relatively constant weight.

It sounds as though leptin should be the answer to the search for a magic diet pill. Obesity is a major health issue in all developed countries and there is much work being carried out by the pharmaceutical industry to find an effective weight loss drug. The problem with most calorie-controlled diet regimes is that people who try to follow a diet can become very hungry, making it more difficult to stick to the diet. If leptin suppresses those hunger signals it would be a great aid to weight loss. Leptin is a peptide, so would have to be injected, but that is less invasive than a gastric band or other surgical intervention. The problem is that, when we are very overweight, the adipocytes produce so much leptin that the leptin receptors become desensitized. This is another example of the hormone resistance syndromes we looked at in Chapter 2. In humans at least, obesity could actually be considered a disorder of leptin resistance. It may be that finding a way of restoring the sensitivity of leptin receptors will be a more effective means of bringing about weight loss in obesity.

Leptin is just one of a number of hormones that work to suppress appetite. As well as regulating blood glucose levels, insulin acts on the hypothalamus to stop us feeling hungry. Several peptides made by the intestine also suppress hunger. They are released by different parts of the gut, in response to the presence of certain types of food in the gut. Glucagon-like peptide (GLP) is a good example. It is released from cells throughout the small and large intestine after a meal, especially a meal containing fats, proteins, and fibre. GLP indirectly suppresses appetite by increasing insulin secretion by the pancreatic beta cells, working together with glucose. GLP also slows down the rate at which the stomach empties, and slows the speed of peristalsis in the gut. This slowing of the gut increases feelings of fullness and means that we feel full and stop eating. The effects of GLP on insulin secretion make it a potentially important target for treatment of type 2 diabetes. Studies so far suggest that it might be useful, both in terms of increasing insulin production and decreasing appetite.

Oxytocin, as we have seen, has a range of behavioural effects. It appears that this peptide also acts as an appetite suppressant. In the region of the hypothalamus that controls appetite, the arcuate nucleus, there is a group of neurons that uses oxytocin as a neurotransmitter. But how does oxytocin affect appetite?

As is often the case, there is much to be learned from a rare genetic disorder. Prader–Willi syndrome is a serious inherited disorder associated with either the deletion or duplication of genes on chromosome 15. People with this condition are unable to regulate their food intake; they are constantly hungry and have an intensely strong drive to eat, leading to severe obesity. In these people there is evidence that the set of oxytocin neurons in the hypothalamus is either missing or not working properly, so there is a clear possibility that lack of oxytocin is involved in the overeating behaviour in Prader–Willi syndrome. These patients have opened our eyes to some of the normal appetite control mechanisms in our bodies which we take for granted.

There is now a lot of evidence that oxytocin suppresses appetite. Injecting oxytocin into the brain of rats stops them eating. However, a lack of oxytocin does not necessarily cause problems with controlling eating behaviour. Mice which had the oxytocin gene knocked out did not overeat when given fatty food, but showed strong preference for sweet foods, compared to the wild type (control) mice. The oxytocin knockout mice did gain more weight than the wild type mice, but this was probably a result of changes in their metabolism rather than the amount they ate. There are oxytocin receptors in taste buds on the tongue which may work to regulate food preferences. This could be the mechanism which changes food preference towards sweet foods in rats that don't produce oxytocin. A recent study has found that people with obesity have a reduced sense of taste and a preference for sweeter and fattier foods. It will be very interesting to find out whether oxytocin is a mediator of this effect.

If we consider all the findings on the effects of oxytocin together, we can see that this hormone has a number of beneficial effects related to social interaction and eating behaviour (see Figure 5.3). We can increase our oxytocin levels by simply having hugs from a friend or stroking a pet. Partners increase their oxytocin through affectionate touch and having sex. Increased oxytocin improves our levels of trust, strengthens our relationships, and stimulates the reward centre in our brain. As a result, we don't seek reward from sweet foods and our immune function may be improved. On the other hand, if we are socially isolated, lonely, or bereaved we have lower oxytocin, making us less sociable, with poorer coping mechanisms, more likely to eat sweet foods, and possibly more vulnerable to infection. Maybe oxytocin is the one hormone we should be looking to increase in all our lives.

Figure 5.3 Some of the known actions of oxytocin on the brain and body

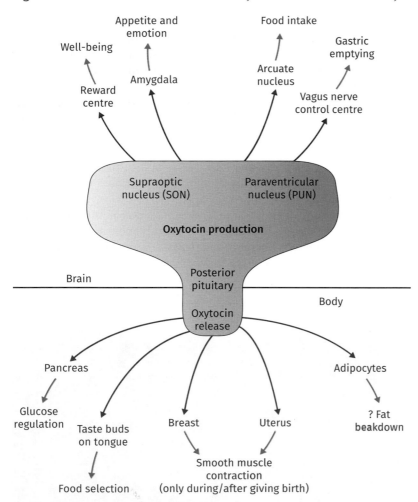

5.3   Testosterone and behaviour

In Chapter 3 we saw how testosterone affects the physical development of babies during pregnancy. Testosterone also affects the development of other mammals and birds. Its effects are especially noticeable in species that show **sexual dimorphism**—in other words, where there is an obvious difference between males and females. We saw in Chapter 1, in the earliest endocrine experiments on cockerels, that removal of the testes (which was effectively removal of testosterone) resulted in a failure to grow the bright red coxcomb which is the main visible difference between hens and cockerels. These castrated cockerels also behaved very differently, not crowing and showing far less aggression than their intact companions.

Other animals also show sexual dimorphism. Male birds of paradise, for example, have brightly coloured plumage with an assortment of extravagant feathers, which they use in territorial displays. Both the plumage and the behaviour are the result of the actions of testosterone. Red deer stags, and the males of several other related species of deer (see Figure 5.4), grow huge antlers every year, which reach their maximum size during the autumn rutting season. This is the time of year when the males compete with each other for the right to mate with the females. The antlers are then shed and regrow the next year. Scientists have shown that blocking testosterone production during the summer causes the antlers to be cast off early, while giving additional testosterone increases antler growth. During the rutting season the stags face each other, taking up a strong position. They often adorn their antlers with vegetation, leaves, moss, and grass. The stags bellow at one another and stamp their hooves on the ground. They charge towards one another and lock antlers in a battle of strength until the weaker male gives way to the stronger. This behaviour is all under the influence of testosterone.

In songbirds, such as the canary (*Serinus canaria domestica*, Figure 5.5), testosterone has several different effects on the brain. It stimulates the male birds to sing more frequently and it alters the quality of their song to make it more attractive to females. It also increases the sex drive and makes

Figure 5.4 Sexual dimorphism in sika deer

Erni/Shutterstock.com

Figure 5.5 A domestic canary. Canary song is influenced by testosterone. This steroid makes male canaries sing more often, using a song that will attract females.

Eric Isselee/Shutterstock.com

males mate more frequently. Using injections of testosterone into precise regions of the brain, scientists have shown that these effects are all independent of each other, depending on which part of the brain is activated by the testosterone.

In male rats and mice, removal of the testes stops them from mating with females, which isn't surprising. As we would expect, giving testosterone to castrated male rats restores their mating behaviour. So what happens if testosterone is given to female rats? Do they start to demonstrate male behaviours? As is often the case in biology, the answer is 'it depends'. Normal female rats injected with testosterone show no change in behaviour; but female rats that have been given testosterone before birth do show male sexual behaviours when given testosterone as adult females, attempting to mount other female rats, for example. It seems that there is a critical period during development when the brain is programmed by testosterone. It is this programming that allows the adult brain to respond to testosterone by displaying male behaviour. So the effects of testosterone on the adult rat brain are not just determined by gender, but also by exposure to testosterone during early development before birth. Because the brains of female rats are not usually exposed to testosterone before birth, adult female rats don't respond to testosterone. What implications does this have for humans? We shall go on and look at the effects of testosterone on human behaviour.

Does testosterone make girls act like boys?

Testosterone has a bit of a reputation for causing bad behaviour in young, and in not-so-young, men. The phrase *testosterone fuelled* is used to describe particularly macho behaviour, aggressive driving, fighting, and many other forms of antisocial behaviour. Is this fair and does giving testosterone to girls or women make them behave like men?

Before birth the brain is at its most *plastic*, meaning that its development can be easily affected by external inputs. The hormonal environment before birth is important in brain development. We saw in Chapter 3 how testosterone, produced by the testes of foetal baby boys, affects the growth and development of the genitals. It also has an impact on the brain.

The bigger picture 5.1

How can we ethically investigate the effects of steroids on the brain?

We have known for many years that there is a measurable difference between a *typical male brain* and a *typical female brain*. The **male brain** is better on average at tasks requiring spatial skills and is less good at verbal tasks and tasks requiring empathy, such as the RMET previously described. The opposite is true for the *female brain*, which is generally better at verbal skills and empathy and less good at spatial skills. Of course, there are very few people who have a totally *male brain* or a totally *female brain*, because there are gradations between the two and most people are somewhere either side of a midpoint. Like any generalization, this tells us nothing about an individual's abilities; some women have exceptional spatial skills and some men have exceptional verbal skills. Nevertheless, it is a measurable fact that, on average, there are gender differences between men's and women's brains. This is not a phenomenon that is unique to humans, and scientists are keen to establish how much of the difference between the two *typical* brain types is due to exposure of the developing brain to androgens before birth.

From an ethical viewpoint it is challenging to carry out experiments to investigate the effects of hormones on human behaviour. We cannot simply inject hormones into the human brain in the same way that we can with fish. Steroids, such as testosterone, can cross the blood–brain barrier and so can be administered as injections into muscle. This approach has been used to study the effects of testosterone on adult men, who have relatively high levels of testosterone anyway, and on adult women, who usually do not. All studies on people have to be considered by an ethics committee, which decides whether the study itself is ethical and if it is being conducted ethically. The people who take part in these studies cannot be coerced into participation and must be free to walk away from the study at any time. The study must be properly explained to them and any risks described. They are then asked to sign a consent form to say that they have understood what the study involves and they are prepared to take part. As you can imagine, there are many more ethical issues which make it almost impossible to carry out studies on foetuses and children; the stages of life when the effects of hormones on the brain are likely to be greatest. Researchers have to approach this question from a different angle.

One approach is to look at people who naturally have higher levels of testosterone than most people of their sex. An example of a naturally occurring condition in which baby girls have high testosterone levels is congenital adrenal hyperplasia (CAH). We saw in Chapter 2 that this condition is caused by a problem with adrenal steroid synthesis, which means that babies produce low amounts of cortisol but high levels of testosterone. The problem in the steroid synthesis can be mild or more severe, with the more severe form resulting in higher levels of androgen production than the mild form. This happens during foetal development, when the brain is developing its hard wiring (making connections between different parts of the brain that remain unchanged into adulthood). Does this exposure to high androgen levels before birth affect the behaviour of the girls? It certainly appears to do so.

Several studies have looked at play behaviour of children with CAH. We know that a certain amount of play behaviour is hard wired in the brain during development and there are gender differences. Typically, boys and girls show a preference for different types of toys and styles of play. Young girls commonly choose to play with dolls and prefer to play *house* and dressing-up games. Young boys commonly choose to play with toy cars or planes and prefer rough-and-tumble play. Some people think these differences are social constructs that arise as a result of the expectations of the adults around the children. However, play differences between the sexes are commonly observed in young mammals of many different species. For example colt (male) and filly (female) foals show different play strategies (see Figure A). Fillies spend more time running, jumping, and bucking, while colts engage in more play that mimics combat and mounting. Female foals also spend about twice as much time in mutual grooming as males, and will groom both sexes, while colt foals almost exclusively groom fillies.

Figure A Colts and fillies play and behave differently from very early in life—without adult expectations

Human studies found that girls with CAH were more likely than other girls to choose the toy cars and planes and rough-and-tumble play, and less likely to choose dolls and dressing-up games. Researchers also found a correlation between the severity of the CAH and play preferences; those girls with the milder form of the disorder were less likely to choose typical male toys than the girls with the severe form. It seems reasonable to conclude that the gender difference in play preferences could be due to exposure of the foetal brain to androgens. The more androgen that the brain is exposed to during development, the more likely it is that a girl will develop more typical *male* play preferences. This is another example of a dose-dependent effect. Another apparent effect of this 'masculinization' is that in tests of spatial awareness girls with CAH are likely to perform better than girls who don't have CAH.

One of the other findings of these studies is that when girls with CAH become adults they are more likely than other women to show a sexual preference for women. The majority of women with CAH identify as heterosexual, but it appears that they are statistically more likely than women in the general population to identify as lesbian or bisexual.

As we noted earlier, there is a spectrum of differences between a typical *male brain* and a typical *female brain*. There are also differences in the amount of androgen that each foetus is exposed to, not just the big differences seen in CAH, but on a small scale in the general population. About twenty years ago scientists suggested that an indirect (proxy) marker of foetal androgen exposure could be the ratio of the lengths of two fingers, the index finger and the fourth (sometimes called the ring) finger. This is called the 2D:4D ratio (see Figure B). Men usually have a slightly longer fourth finger than index finger, giving a low 2D:4D ratio. In women the two fingers are more equal in length or the index finger is a little longer than the ring finger, giving a higher 2D:4D ratio. It seems that this ratio is determined by the amount of pre-natal androgen exposure, which may also determine how *male* the brain becomes. Since the start of the twenty-first century many studies have used this simple measure to correlate behaviours with the degree of masculinization of the brain.

Figure B The 2D:4D ratio has been shown to be a good proxy marker of foetal androgen exposure

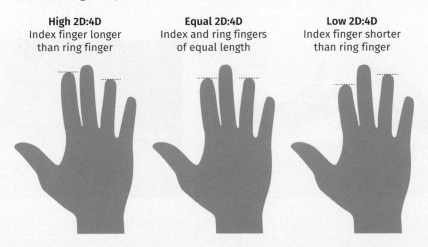

High 2D:4D
Index finger longer
than ring finger

Equal 2D:4D
Index and ring fingers
of equal length

Low 2D:4D
Index finger shorter
than ring finger

We started this section by asking how you can ethically investigate the effects of steroids on the brain. Few people would guess that the ratio of finger lengths could be used as a proxy measure of prenatal androgen exposure. Scientists often need to use indirect measures like this in human studies. Obviously, this method has limitations, but well-designed studies using this measure have increased our understanding of the basic science involved. It is also worth bearing in mind that all research methods have limitations and there is no such thing as the *perfect study* in biology.

❓ Pause for thought

- Are there other aspects of health or behaviour that might also be influenced by androgen exposure before birth? Do some research to find out more about current models of the impact of sex hormones on behaviour in children and adults.
- We have described a number of different experimental approaches in this chapter. Think about what the limitations of each method might be.

While it appears that testosterone can affect the developing brain to influence behaviour, what effect does it have on adult women? There have been several studies where testosterone or a placebo (inactive substance) have been given to women and different aspects of behaviour have been observed. A single dose of testosterone can improve the spatial skills of young women in a virtual reality game, although it does not affect performance at learned tasks like navigation, which is rather more complex. Testosterone administration also lowers the scores of women doing the RMET. Other studies have shown that women given testosterone treatment perform more like men when given a range of different behavioural tasks, including games of trust such as the Prisoner's Dilemma. In these studies, testosterone has been found to decrease trust and increase risk-taking behaviour. However, before using testosterone—or the lack of it—as an excuse or explanation for perceived strengths or weaknesses in either sex, you also need to know that women who have been given a placebo but told that they have been given testosterone, show exactly the same increase in risk-taking behaviours as those women who were really given testosterone! This is a good example of why complex behaviours are so difficult to study and at the same time so fascinating to researchers.

Interestingly, not all research has found the same results, and there have been studies that have shown no effect at all of testosterone on women's behaviour. It all looked a bit of a muddle until researchers started to factor in the question of pre-natal androgen exposure. Scientists looked at the effects of testosterone on scores in the RMET. Taken as a whole group the testosterone had a tiny overall effect, slightly lowering the scores. However, by measuring 2D:4D ratios scientists could separate women into two groups: high and low prenatal androgen exposure. The women with low

Figure 5.6 This graph shows the effect of testosterone on womens' performance in the RMET. The impact of testosterone on the adult female brain appears to be strongly affected by the level of exposure of her brain to testosterone during foetal development. Values shown are mean +/- standard error of eight women in each group.

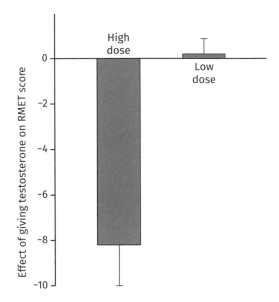

Data taken from van Honk et al. (2011). *PNAS*, 108, 3448–52.

prenatal androgen exposure had no response at all to the testosterone; their scores were completely unchanged in the RMET. However, the women who had high prenatal androgen exposure showed a strong response to testosterone, with significantly lower scores after treatment. It looks as though the human brain is primed by androgens before birth, which makes the brain sensitive to the effects of androgens in adult life (see Figure 5.6). This is exactly the same effect that we saw in the studies on rats.

Effects of behaviour on testosterone levels

So far in this section we have looked at how testosterone can affect behaviour. We now know that behaviour can affect testosterone levels too; particularly competitive behaviour. A number of studies have looked at the effects of playing football on blood testosterone levels. Scientists took blood samples from both male and female football players before and after their matches. Playing football has a significant effect on the levels of testosterone in players' blood, but the effect depends on whether their team won or lost the match. The winners had an increase in testosterone while the losers showed a decrease. The same effect was seen in both men and women. There is even some evidence that the same effect is seen in sports fans; increased testosterone when their team wins and decreased testosterone when their team loses.

As we discussed at the start of this section, some animals live in family groups which have a social structure or *pecking order*, with a dominant male at the top of the group. It is often only these dominant individuals that are sexually active and they try to prevent any other males from mating with the females. It has been known for many years that testosterone is the key to becoming the dominant male in a group. Many studies have shown that males that rise through the social hierarchy and become dominant have higher testosterone levels than other males in the group. It was initially thought that the increase in testosterone came first and that this caused increased aggression and a better outcome from confrontation between males. Scientists working with an African cichlid fish, *Astatotilapia burtoni*, have shown that it is the change in social status that causes the change in testosterone levels, not the other way round. In a group of these fish, when a subordinate male becomes dominant, for example as a result of the death of the group's dominant male, the newly dominant male undergoes a rapid increase in their plasma testosterone levels.

In male deer, it is contact with female deer which causes an increase in both testosterone levels and in sperm count. It follows that males which are driven away from a herd by dominant males will go on to have lower testosterone levels and be less capable of successfully mating.

In this chapter we have looked at the effects of just a few hormones on behaviour. We have looked at some of the methods used to study behavioural effects of hormones and considered the issues around ethics and suitable study methods. There are, of course, far more hormones that act on the brain than those we have described here. Chapter 6 is about the use and abuse of hormones. There are examples of effects of different hormones on mood and on behaviour. Keep an eye out for them.

Case study 5.2
Spotted hyenas

In mammals, as we saw earlier, it is commonly the males who are the larger and more aggressive of the sexes. This is particularly seen in social mammals where the males use their size and aggression to compete for dominance over the group and to win the right to mate with the most females. Often the difference between males and females is clearly visible, with males having adaptations which are used to establish dominance. There are a few species of mammals where the opposite is true, and it is the females that are larger, more aggressive, and dominant in the group. One especially interesting example is the African spotted hyena (*Crocuta crocuta*). The spotted hyena is a type of hunting dog which lives in social groups in sub-Saharan Africa (see Figure A). These social groups are unusual in that they are competitive, not cooperative. Females will fight each other, as will males, and the females provide food only for their own cubs and don't share within the wider group, which is unusual for a pack animal.

Figure A The spotted hyena lives in social groups which are led by dominant females. This social structure is associated with high levels of testosterone exposure to females before birth.

Sergey Uryadnikov/Shutterstock.com

In the spotted hyena the placenta produces unusually large amounts of androgens. This means that all the developing cubs, both male and female, are exposed to high levels of testosterone before birth. This has an effect on the development of the female cubs. These cubs have genitals which are quite different from most other female mammals. Most female mammals have separate openings for urination and for reproduction; typically called the urethra and vagina. In addition, female mammals have a clitoris, a small button of very sensitive tissue which can become erect. In women it is usually the main site of sexual pleasure. Male mammals have a combined urinary and genital opening in the erectile tissue called the penis. The female spotted hyena has an anatomy that is much more like the male arrangement. They have a very large clitoris, which can become erect, like a penis, and which holds both the urinary and reproductive openings. So female hyenas urinate, mate, and give birth through their clitoris.

We have seen how exposure to androgens before birth can programme the brains of rats and humans to respond to androgens in adult life. The hyena is a particularly extreme example. It seems that the very high androgen levels before birth cause masculinization of the brain of the developing cubs. There is a correlation between the amount of androgen and the behaviour of the cubs. The most dominant females in the social group produce more androgen in their placenta than subordinate females. Their cubs are far more aggressive than the cubs of subordinate females. The cubs all engage in play fighting but the cubs exposed to the highest androgens show the most aggression.

In other species, as we saw earlier, prenatal testosterone exposure primes the brain to respond to androgens in adulthood, raising the question of testosterone levels in adult hyenas. Perhaps surprisingly, the levels of testosterone are low in adult female hyenas; over twenty times lower than in males. However, scientists found a different androgen, called **androstenedione** (see Chapter 1, Figure 1.8) that is equally high in both male and female hyenas. This steroid can easily be converted into testosterone in different tissues of the body, such as the brain. So it is quite possible that the adult hyena brain is exposed to much higher effective levels of testosterone than the blood levels of the hormone would suggest.

❓ Pause for thought

What advantage does it give hyenas to have such an unusual female phenotype as a result of high prenatal androgens?

⬚ Chapter summary

- The mammalian hypothalamic hormones, oxytocin and vasopressin, have counterparts in all vertebrates. All of these variations appear to have originated from a single gene, duplicated and copied forwards or backwards, and with individual changes in some amino acid positions.
- Many hormones have effects on behaviour in addition to their better-known effects on the body.
- Oxytocin affects a variety of complex behaviours, including maternal bonding, attachment, trust, empathy, and eating behaviour.
- Several different hormones interact to control appetite and food intake: ghrelin stimulates hunger and increases food intake; leptin inhibits hunger and regulates fat stores; GLP, insulin, and oxytocin all work to stop you from eating when you are full. Oxytocin also works to control food preferences.
- Testosterone determines male behaviour. The absence of testosterone determines female behaviour. Masculinizing effects of testosterone on female animals are only observed when the female brain has been *primed* by exposure to androgens before birth.
- Winning or losing a sports game seems to affect the testosterone levels of both players and spectators. We are finding an increasing number of other examples where behaviour affects testosterone levels in vertebrates.

⬚ Further reading

https://www.ncbi.nlm.nih.gov/books/NBK278953/
For a detailed account of congenital adrenal hyperplasia.

Slimani, M., Baker, J.S., Cheour, F., Taylor, L., and Bragazzi, N.L. Steroid hormones and psychological responses to soccer matches: insights from a systematic review and meta-analysis. (2017). *PLoS One* (2012) 12(10): e0186100.

https://vetfolio-vetstreet.s3.amazonaws.com/mmah/b6/da39af1779477 ca4bf98ff34b687a9/filePVE_02_11_318_1.pdf
Tells you more about the behaviour of foals. Social transitions cause rapid behavioral and neuroendocrine changes.

Maruska, K.P. (2015). Social transitions cause rapid bevioural and neuroendocrine changes. *Integr. Comp. Biol.* 55(2): 294–306. doi: 10.1093/icb/icv057. Epub 2015 Jun 1. Review.
A review of the work on cichlid fish.

http://www.pnas.org/content/pnas/early/2013/11/05/1311371110.full.pdf
The studies on canary song are in this paper.

Domes, G., Heinrichs, m., and Herpetz, S. (2014). 'Oxytocin promotes facial emotion recognition and amygdala', *Neuropsychopharmacology* 39 (3), 698–706.

Lee, S.Y., Lee, A.R., Wangh, R. et al. (2015). 'Is oxytocin application for autism spectrum disorder evidence-based?', *Experimental Neurobiology* 24 (4), 312–24.

This link shows you how to take part in the Harvard RMET study and test your emotional intelligence: http://socialintelligence.labinthewild.org

Discussion questions

5.1 How is it that the same hormone has evolved to have totally different effects in different species?

5.2 Given the importance of oxytocin in relationships and trust, is it acceptable to manipulate hormone levels in order to change behaviours?

5.3 Which hormones would you target if you were designing an anti-obesity pill? Would the introduction of such a pill have any ethical considerations?

5.4 The origins of human behaviour are complex, being affected by social constructs in addition to biology. How would you set out to explore the relative contribution of social inputs and biology to the observed differences between the play behaviours of boys and girls?

5.5 How can you tell the difference between a view based on scientific study and a belief?

6 THE USE AND ABUSE OF HORMONES

Many hormones are used as medicines (see Figure 6.1). We saw in Chapter 2, when we looked at autoimmune damage to the pancreas, thyroid, and adrenal glands, that disease can prevent the body from making some hormones. In these situations, we need to replace these hormones in the body. Some-

Figure 6.1 Hormones are used as medicines for many reasons. Some hormones, such as steroid and thyroid hormone can be taken as tablets as they are absorbed unchanged from the gut. Large peptide hormones, such as insulin, need to be injected. Small peptide hormones, like vasopressin, can be given by nasal spray because it can be absorbed directly across the nasal mucosa.

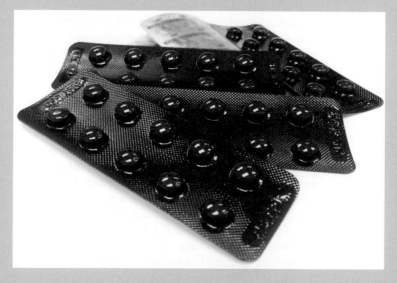

Arpon Pongkasetkam/Shutterstock.com

body who has type 1 diabetes is unable to produce insulin, and so needs to receive insulin replacement therapy. Insulin, as a large peptide hormone, is administered as an injection several times each day. Both thyroid and adrenal hormones can be given as tablets. As well as this very straightforward use of hormones as replacement therapy, there are situations where hormones are used as medicines. In Chapter 4 we looked at the use of steroid hormones as contraceptive agents. We also use hormones to treat illnesses, such as the use of adrenal steroids as anti-inflammatory drugs to treat asthma. There are also times when hormones are used to gain an unfair advantage in sports or as unauthorized diet pills. In these situations, hormones become drugs of abuse.

6.1   Hormone replacement therapy

When an endocrine gland is unable to make enough of a hormone to maintain homeostasis, it is usually necessary to replace that hormone. The two hormones that most commonly need to be replaced are thyroxine and insulin. This is because the thyroid gland and the islets of Langerhans in the endocrine pancreas are the two tissues most likely to be damaged by autoimmune endocrine disease, as we saw in Chapter 2. Technically, the replacement of any hormone could be called hormone replacement therapy, but in practice this term has come to mean something very specific indeed: the replacement of oestrogens in women at the time of the menopause.

A woman's reproductive life starts with menarche, the onset of monthly menstrual periods, and ends with **menopause**, when periods stop and a woman is no longer fertile. This happens at around the age of fifty or fifty-one years. Unlike menarche, the age at which a woman enters the menopause has not changed over the past 150 years. Menopause is not usually a single event, but a process that happens over a time course of several years. The remaining follicles in the ovary become less able to respond to follicle stimulating hormone (FSH), so ovulation becomes irregular and then stops, which means that menstruation also becomes irregular and then stops. As the ovary isn't responding to hormonal stimulation, levels of the sex steroids, oestradiol and progesterone, also decrease.

When her levels of oestrogens fall a woman can experience a number of symptoms, which are the classical symptoms of the menopause. These include hot flushes, difficulty sleeping, loss of sex drive, and low mood, which can all be a result of a decrease in oestrogen levels. These are the changes that a woman might be aware of, but there are other effects which are not visible, especially effects on bone. Without oestrogens, bones gradually lose part of their mineral content and the bone mineral density falls. This can lead to a condition of bone weakness called **osteoporosis**, which results in an increased risk of fractures.

Not all women experience symptoms of menopause. First, there is a lot of natural variability between different individuals, and secondly there is a strong cultural aspect to how the menopause is viewed. In cultures where older people are highly respected and valued, women do not generally experience problematic symptoms of menopause. However, in countries where youth and beauty are valued above age and wisdom, the end of a woman's reproductive life can be viewed as the end of beauty and desirability. Women in these societies are significantly more likely to have a negative experience of menopause.

In developed countries it is quite usual for women to take hormone replacement therapy (HRT) to treat menopausal symptoms. HRT is very effective at preventing hot flushes, restoring sex drive, and preventing bone mineral loss. HRT usually consists of a daily tablet containing a combination of an oestrogen and progesterone. Although taking an oestrogen on its own is very effective, it is always combined with progesterone. This is because oestrogens promote cell proliferation, as we saw in Chapter 4. In the first half of the menstrual cycle, when oestrogen levels are much higher than progesterone, the lining of the uterus grows and develops a blood supply. This is a direct result of the actions of oestrogens in the absence of progesterone. After menopause this effect can result in cancer of the lining of the uterus, called endometrial cancer. Adding progesterone to the HRT tablets stops oestrogens from causing proliferation of tissues and reduces the cancer risk. No medicine is entirely without risk and long-term use of HRT seems to cause a slight increase in the risk of developing breast cancer and may increase heart and some circulatory problems.

Many women take the view that menopause is a natural process and don't like the idea of taking hormone tablets. They prefer a *natural* remedy and so change their diet to control menopausal symptoms. There are several different types of plant that naturally contain a form of oestrogen. The best known of these is the soya bean (see Figure 6.2), which contains an oestrogen called genistein. Because of this, soy products are often marketed as a natural therapy for menopausal symptoms.

Figure 6.2 Many women are happy to use HRT prescribed by their doctors, but natural products from soy contain phytoestrogens which also help some people to overcome their symptoms

naito29/Shutterstock.com

Scientific approach 6.1
Making hormones to use as drugs

Hormone deficiencies could be treated effectively long before scientists had even identified hormones. As early as the fifteenth century a Chinese physician treated thyroid deficiency with minced thyroid glands, for example. In the West, it wasn't until 1891 that George Murray, an English physician, began using an extract of thyroid glands to treat the symptoms of an underactive thyroid. This approach had a number of problems, not least of which was the question of where to get the thyroid gland, from animals or from dead people. George Murray used fresh sheep's thyroid, obtained from abattoirs. Similarly, extracts of pig or cow pancreas were routinely used to treat type I diabetes from 1922 until the 1980s. At first, the very impure extracts of animal pancreas caused an allergic response in about half of patients, mostly at injection sites. Over time the extracts became increasingly pure but the problem with using a peptide hormone obtained from another species to treat a human is that the amino acid sequences are not always the same. Pig insulin is very similar to human insulin, with only one amino acid difference, but cow insulin has three different amino acids (see Figure A). The different forms of insulin all bind to the human insulin receptor and cause a very similar response to human insulin.

Figure A As you can see, there are small differences in the amino acid sequence of human, cow, and pig insulin. In spite of these differences, all three forms of insulin can activate the human insulin receptor.

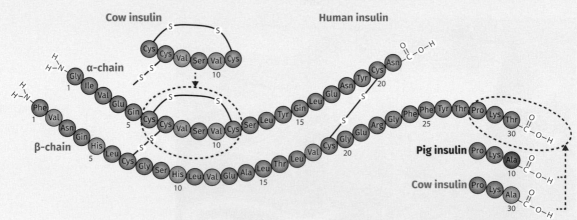

Unfortunately, even the highly purified pig or cow insulin (see Figure B) was recognized as 'foreign' in a small proportion of patients and triggered an immune response. When this happened, the person started to produce anti-bodies to the animal insulin, which bound to the insulin and stopped it from attaching to the receptor. The animal insulin was effectively neutralized. Only about 3% of people treated with animal insulin had an immune response, but it caused them real problems.

This isn't an issue for steroid hormones or thyroid hormones because they are not peptides and have an absolutely identical structure in every animal. Unlike peptide hormones, steroids are never recognized as 'foreign' by the body.

Figure B It's not really surprising that hormones from cows caused problems for some people—but for many years, animal-based insulin was the only way to treat individuals affected by type 1 diabetes

aleks.k/Shutterstock.com

For some peptide hormones the amino acid sequence is so different that there isn't an animal version that is effective in humans. The obvious solution to this problem is to use endocrine tissue obtained from humans to extract and purify the required hormones. Growth hormone (see Chapter 1, Table 1.1) is a good example of a hormone which had to be obtained from human tissue because the animal versions of it simply don't work in humans. Growth hormone deficiency was a very difficult condition to treat. Extracts of animal pituitary glands had no effect and so in the 1950s scientists started purifying growth hormone from human pituitary glands (obtained after the donors had died!). Children with severe growth hormone deficiency were treated very effectively with these extracts until the mid 1980s.

At this time there were several reports of people who had been treated with human pituitary extract developing a prion disease called Creutzfeld

Jacob disease (CJD). This progressive and fatal brain disease was rapidly linked to the use of pituitary extracts and so growth hormone derived from human tissue was immediately withdrawn from use.

Fortunately, advances in scientific knowledge had progressed far enough to allow the rapid development of an alternative source of growth hormone. The gene which codes for growth hormone had been sequenced and cloned in 1979 and the first recombinant human growth hormone was produced in 1981 (see Figure C). Recombinant technology snips the DNA sequence for a protein out of the chromosome and inserts it into the genome of a micro-organism, such as *E. coli* or a yeast. The micro-organism can be grown in large quantities and the protein purified. For a single-chain peptide such as growth hormone this is a reasonably straightforward process. By the time the pituitary extracts were withdrawn, recombinant human growth hormone was already available as a replacement.

Figure C Genetically modified micro-organisms have revolutionized peptide hormone treatments and guarantee a regular, pure supply

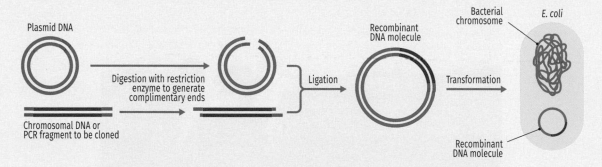

Reproduced from Craig et al, *Molecular Biology* 2/e. By permission of Oxford University Press

All the peptide hormones used therapeutically today, including insulin and growth hormone, are made by recombinant technology using human DNA. This gets over all the problems of extracting hormones from either animal or human tissues. Making steroid and thyroid hormones is a much more straightforward process. As these are small and relatively simple molecules, they are synthesized and purified in laboratories using chemical processes.

❓ Pause for thought

There were ethical considerations around the use of hormones extracted from animal or human tissues; both vegans and people with certain religious views had difficulties accepting the use of these hormones. Do you think that there are any ethical concerns around the use of hormones made by recombinant technologies?

6.2   Hormones as anti-inflammatory drugs

As recently as the 1940s there was no effective treatment for rheumatoid arthritis, a long-term progressively worsening illness which causes painful swollen joints and can severely limit your activity. It was first thought that this disease was due to infection, but by the 1940s it was recognized to be a result of chronic inflammation. The discovery that steroids extracted from an adrenal gland had anti-inflammatory properties was made around the same time. The steroids came from the outer part of the adrenal gland, the adrenal cortex, and so were called *corticosteroids*. The first patient to receive corticosteroid therapy for their rheumatoid arthritis was treated in 1948 by Dr Philip Hench, who later received a Nobel prize for his work. Corticosteroids were hailed as a 'miracle cure' for rheumatoid arthritis and other inflammatory conditions and were widely used in the 1950s.

From the start it was apparent that there were adverse effects of corticosteroid use. Dr Hench's first patient had *facial puffiness* and reported a dramatically altered mood, becoming hostile and depressed. Other patients had similar symptoms and also developed high blood pressure and a fast heart rate. The first warning about long-term corticosteroid treatment was published in 1951. There were particular concerns about what happened when somebody stopped taking corticosteroids after a long course of treatment; the effects were the same as adrenal failure (see Chapter 2). In 1960 guidance for steroid use was published, advising doctors that corticosteroids should *never* be the first course of treatment for arthritis or other inflammatory disease. In spite of the adverse effects (see Figure 6.3), the simple fact was that corticosteroids worked; they reduced pain and swelling in rheumatoid arthritis and generally caused a dramatic improvement in the inflammatory condition. The doses of steroids being used were very high indeed and they were being given systemically. This means that the patient was getting injections or taking tablets so that their whole body was exposed to the steroid, not just the part that was affected.

We now have a good understanding of how corticosteroids have their effects on the immune system to reduce joint pain and swelling. Corticosteroids work by blocking the production of cytokines, the chemicals that cause the symptoms of inflammation. They act to increase the expression of some genes (see Chapter 1) and decrease the expression of others, and in some cases they work by increasing the breakdown of certain species of mRNA, so blocking translation rather than affecting gene expression. It is this wide range of mechanisms of action and the effect of corticosteroids on many different genes which helps to make them particularly effective.

Although corticosteroids block the signalling pathways leading to inflammation, they don't prevent whatever caused inflammation in the first place. So they can treat the *symptoms* of a disease but can not alter the underlying *cause* of that disease. Modern medicine has come a long way since 1960 and we now have a group of drugs called non-steroidal anti-inflammatory drugs (NSAIDS), such as ibuprofen, which can be used to treat inflammation. Diseases like rheumatoid arthritis, which are autoimmune diseases, are now treated with immune suppressants, which address the underlying cause of the disease and so are called disease-modifying drugs.

Figure 6.3 The long-term use of oral corticosteroids causes a range of serious adverse effects, as shown here. This condition is known as Cushing syndrome.

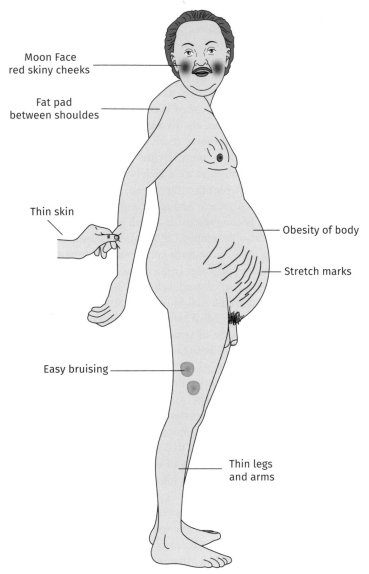

Moon Face
red skiny cheeks

Fat pad
between shouldes

Thin skin

Obesity of body

Stretch marks

Easy bruising

Thin legs
and arms

Corticosteroids are still used widely as anti-inflammatory drugs, but they are used in very different ways to those early studies. If you look in your bathroom cabinet you might find a tube of corticosteroid cream. If you have asthma, then you might use inhaled steroids to prevent attacks. The corticosteroids most commonly used are hydrocortisone (which is exactly the same as the corticosteroid hormone cortisol), beclomethasone, and prednisolone (see Figure 6.4). They are now mostly used by being directed

Figure 6.4 Corticosteroids and synthetic versions of these hormones, used correctly, can dramatically improve the life of people suffering from conditions ranging from rheumatoid arthritis to asthma and eczema. This image shows the hormone cortisol and two synthetic versions of cortisol, beclomethasone and prednisolone. They all act through intracellular receptors, as shown in Figure 1.13.

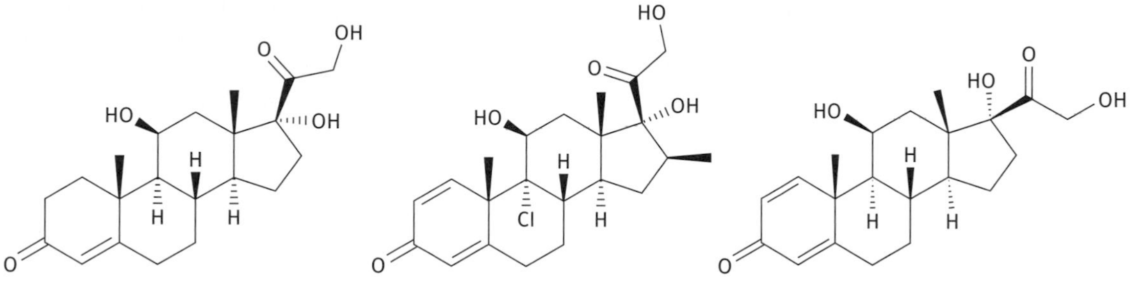

(a) Cortisol   (b) Beclomethasone   (c) Prednisolone

specifically to the part of the body that is affected by the inflammation. Corticosteroid creams are used to treat skin inflammation, for example as a result of eczema. In the same way, beclomethasone is used in an inhaler to deliver the steroid directly into the airways, to reduce the sensitivity to triggers of asthma. Sometimes rheumatoid arthritis is treated with a corticosteroid injection directly into the affected joint. By using steroids in these ways it is possible to use much smaller amounts of the steroid to obtain the effect that is needed. It also means that the effects of the steroids are concentrated locally, in one tissue or joint, and only very small amounts get into the bloodstream, reducing the chance of any adverse effects. Steroid tablets are still used sometimes, for example when somebody with rheumatoid arthritis has several joints affected. But oral corticosteroids are only used for a very short time, to control a flare-up of the disease, never for months at a time as was the case in the past.

6.3   Thyroid hormones and weight loss

We saw in Chapter 2 that thyroid hormones have effects on nearly every cell in the body. One of the important effects of these hormones is to increase the basal metabolic rate. People with an overactive thyroid gland have a higher metabolic rate and so tend to lose weight. When the thyroid gland is underactive, then thyroid hormone tablets are taken to make sure that thyroid hormone levels stay within the correct range. With these two facts it is not a huge leap to imagine that taking thyroid hormone tablets with a normal thyroid function might help you to lose weight. Unfortunately, the internet makes abusing thyroid hormones very simple; there are many websites that offer to sell thyroid *supplements* without a prescription or any proper medical assessment. There are also books available, with titles that make a direct link between thyroid hormones and weight loss. None of the websites or books explicitly recommend taking thyroid hormones in order to lose weight, but there is an implication that increasing thyroid activity will help. Very few of the websites actually warn about the dangers of taking extra thyroid hormone.

With the current 'obesity epidemic', should everybody take additional thyroid hormone to help them maintain a healthy weight? The answer to that is *absolutely not*! The effects of taking additional thyroid hormone when you haven't been told to by a doctor can be very unpleasant and dangerous. It can even kill you. Let's look at a case to see what happens when you take thyroid hormones that are not clinically needed.

Case study 6.1

Jude's story

Jude was getting married in five weeks' time. She had been fitted for her wedding dress several months earlier, but she had put on weight and wanted to lose it quickly so that her dress would fit. She was really worried about the amount she had paid for the dress and wanted to look perfect on the big day. Jude's aunt had been diagnosed with an underactive thyroid gland about a year ago. Her aunt had put on a lot of weight before starting to take thyroid hormone tablets, but since starting treatment her weight had come back down. Her aunt's dose of thyroid hormone had been changed a couple of times and Jude found a pack of her aunt's old tablets in the bathroom cupboard. Jude thought that it would help her to lose weight if she tried her aunt's tablets so she started taking two each day; she figured that her aunt had lost weight quite slowly just taking one a day so she needed a bit more. After a week she hadn't lost any weight at all, although she had been getting diarrhoea. She decided that they can't have been strong enough for her and doubled the dose she was taking.

Jude's friends were getting worried about her. She was usually calm and quiet but had become very restless. She was snappy and irritable with them and getting increasingly anxious and upset. She had been complaining about feeling hot and sweaty, not being able to sleep, and a stomach upset, which meant she had to keep going to the toilet. She mentioned that she had been having palpitations and that her heart was going really fast. At first her friends thought it was just wedding nerves, but she started getting very agitated, shaking, and sweating. It was clear that she wasn't well so they took her to the local hospital.

The doctors found that she had a temperature of 40°C, with high blood pressure, and a pulse rate of 180 beats per minute. She was confused and couldn't remember if she had taken any drugs, but her friends said that she didn't do drugs. She was admitted to hospital and given treatment to try to slow her heart rate and lower her blood pressure. Blood tests showed that she had high levels of thyroid hormone and no measurable TSH. Her heart rate was irregular but very fast and so she was transferred to the intensive care unit, where she stayed for two weeks. The wedding had to be cancelled because Jude was still in hospital. The doctors explained that she could easily have died.

The science behind the story

We know that thyroid hormone receptors are found in nearly every cell in the body, so perhaps we should expect raised thyroid hormone levels to have such a wide range of effects. All the symptoms that Jude and her friends noticed were a result of the thyroid hormones: feverishness, diarrhoea, shaking, restlessness, inability to sleep, and irritability. Thyroid hormones also directly affect the heart, causing an increase in heart rate and often making the rhythm irregular. This causes the sense of palpitations that Jude described. You can see from Jude's story that the effects of thyroid hormone aren't seen immediately. The final effect is a bit like taking adrenaline, but there is a much longer gap between taking the hormone and seeing the effects. If we look at how thyroid hormone works on the heart we can understand why this happens. Thyroid hormones increase the effects of adrenaline and noradrenaline in the body. They do this by increasing the numbers of receptors for these hormones, which takes place over a timeframe of hours or days. This means that the body has a much greater response to adrenaline and noradrenaline than usual, resulting in increased heart rate and raised blood pressure. It also means that the heart rate is more likely to be irregular, with a serious risk of heart failure. The increased actions of adrenaline and noradrenaline also explain the restlessness, anxiety, lack of sleep, and diarrhoea. Jude also felt hot and sweaty, which is caused by the increased metabolic rate.

There is no antidote for thyroid hormone so the doctors had to wait for her blood levels to decrease through normal metabolism and excretion. Thyroxine has an unusually long half-life in the body. It takes six to seven days to decrease blood levels by 50%. In Jude's case her hormone levels were so high that it took two weeks for her thyroxine levels to fall sufficiently for her to leave the intensive care unit. She was at high risk of heart failure for the whole of this time.

❓ Pause for thought

Some people think that being overweight indicates that they have a *glandular problem*. If we assume that *glandular* means hormones, is there evidence to support this belief?

6.4   Hormones in sport

All athletes in competitive sports are looking to get an advantage over their opponents, to find that extra boost to their performance which allows them to win. There are many ways to do this fairly, within the rules of the sport, by improving diet or training regimes, but some athletes cheat by taking drugs which are known as *performance enhancers*. Many of the drugs that work as performance enhancers are hormones. The use of hormones as

performance enhancers has been going on for thousands of years. Gladiators in the arenas of ancient Rome would eat bull's testes to improve
their strength and performance—see Figure 6.5. For hundreds of years the
use of hormones in sport remained at this level, but as endocrinology developed as a science in the early and mid-twentieth century, the use of
synthetic hormones and chemical derivatives of hormones became more
widespread. The International Amateur Athletics Federation banned the use
of performance-enhancing drugs as early as 1928. At this time they were
mostly concerned about non-hormonal stimulants such as amphetamines.
The ban was difficult to implement because there was almost no way to test
for drug cheats. Other sports bodies followed through the twentieth century
and in 1999 the World Anti-Doping Authority (WADA) was established by

Figure 6.5 Trying to improve sporting prowess by legal means or buying
illegal substances seems almost as old as sport itself

INTERFOTO / Alamy Stock Photo

the International Olympic Committee. Since the mid-twentieth century there has been an evolving battle between the drug-testing authorities and the sports teams who wish to cheat. New performance-enhancing drugs are constantly being developed by pharmacological scientists and there is constant pressure on the scientists at WADA to develop better methods for detecting these drugs.

The main hormones that are used by competitors are anabolic androgenic steroids, growth hormone, and erythropoietin (EPO). Let's take a look at why these are used and what effect they have.

Anabolic androgenic steroids

This group of steroids consists of a large number of different drugs, all related to the male sex steroid, testosterone. The anabolic androgenic steroids are called anabolic because they cause an increase in growth of body tissues, stimulating uptake of amino acids, and promoting protein synthesis in cells. They are called androgenic because they all act by binding to the androgen receptor, an intracellular receptor which acts as a transcription factor, altering gene expression in target cells (see Chapter 1, Figure 1.13). This is the same receptor that is activated by testosterone and 5 alpha dihydrotestosterone and so anabolic androgenc steroids mimic the effects of the naturally occurring male sex steroids. Testosterone itself is an anabolic androgenic steroid, which is often abused by athletes. However, work by pharmaceutical chemists has led to the development of a wide range of drugs that activate the androgen receptor. Commonly these are called *anabolic steroids*, and that is how we will refer to them.

Use of anabolic steroids to enhance athletic performance goes back centuries, well before testosterone had been named. The logic is very simple; a bull is much bigger, stronger, more muscular, and more aggressive than a cow. If you remove a bull's testes it becomes more docile, less muscular, and less strong. So, in order to become strong, powerful, and aggressive, like a bull, an athlete should consume bull's testes. Simple and apparently logical at the time. This is how anabolic steroid use in sport started. As we saw in Chapter 1 from Berthold's experiments in the nineteenth century, removing the testes from cockerels removed their male characteristics, and replacing them restored these characteristics. Similarly, if men have very low testosterone levels, giving testosterone replacement therapy decreases their body fat and increases muscle mass, making them leaner and more muscular. However, it doesn't necessarily follow that giving additional testosterone, or other anabolic steroid, to men who aren't deficient in the hormone will make any difference to their strength or athletic ability. There is also the question as to whether giving anabolic steroids to women might improve their athletic performance.

Of course, a drug doesn't have to have any measurable effect on muscle or other part of the body in order to increase athletic performance. Many sports have a strong psychological component where self-belief is almost as important as diet and exercise regime. So, if an athlete *believes* that a drug will improve their performance, it almost certainly will. To investigate in a research study whether a steroid affects muscle strength it is important that the athlete doesn't know that they are taking the steroid, because the knowledge can affect their performance. To overcome this problem, scientists use an inactive treatment, called a placebo control. They compare the effects of the placebo to the effects of the actual treatment and measure the difference.

Studies looking at the effects of testosterone have shown some clear effects. For example, one study by a group of doctors and scientists from the USA divided forty-three healthy men into two groups. One group was given an exercise regime to follow, the other group was not. Half of each group were given testosterone and the other half had placebo injections. The treatment lasted ten weeks and the men had a series of tests at the start and end of the study. The researchers looked at a number of metabolic and biochemical outcomes, but the most interesting results are the effects on muscle size and strength. Some of the results are shown in Figure 6.6. As you can see, exercise significantly increased muscle bulk and strength. Testosterone increased both these measures without any exercise, and significantly enhanced the effects of exercise.

As testosterone is so effective at improving muscle bulk and strength, even in healthy men, there is an argument that its use in sport should be permitted. It has been suggested that this would eliminate the problem of

Figure 6.6 The effects of testosterone treatment, compared with placebo control in normal men

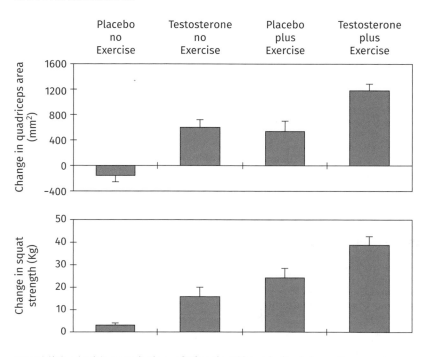

Republished with permission of Bioscientifica Limited, from *Hormones and Sport, Proof of the effect of testosterone on skeletal muscle*, S. Bhasin, L. Woodhouse, and T.W. Storer (2001); permission conveyed through Copyright Clearance Center, Inc.

cheating; everybody would have the option to use steroids and so there would be a level playing field. The problem is that there are adverse effects of steroid use which make the issue more ethically complex. In Chapter 4 we looked at how testosterone treatment could be used as a form of male contraceptive. Taking anabolic steroids at even low doses can make a man infertile. At the much higher doses used to increase athletic performance there are additional risks. These steroids cause a change in lipid and glucose metabolism, causing insulin resistance, and significantly increasing the risk of heart attack. They also cause growth of the prostate gland and may increase the risk of prostate cancer. In people with certain existing mental health problems anabolic steroids are thought to cause an increase in angry behaviour: so-called 'roid rage'. Long-term use also increases the risk of liver cancer, especially when steroid tablets are taken. By far the most common adverse effect, however, which is seen even at relatively low doses and in short-term use, is acne. Anabolic steroids cause really bad acne. Taken together, steroid abuse is something that sports authorities wish to discourage (see Figure 6.7). The choice to use anabolic steroids is more than simply a decision to cheat, it is a question of what level of health risk an athlete is prepared to tolerate in order to improve their chance of winning.

Figure 6.7 Ben Johnson 'won' the 100 m gold medal in the Seoul Olympics in 1988, setting a new world record, and leaving his fellow athletes trailing behind him. But within two days he had been stripped of his medal, after his urine tested positive for an anabolic steroid. He was much faster than his rivals because he cheated, not because he was a better runner. This was a significant event in Olympic history and resulted in the introduction of closer monitoring of all athletes. It also changed public perception so that successful athletes came to be viewed with suspicion.

Photo by Leo Mason/Popperfoto/Getty Images

It is not only in professional sports that anabolic steroids are abused. In gyms, especially those with a focus on bodybuilding, and in some amateur sports, anabolic steroids have been part of the culture for many years. That is not to say that everybody who works out at a gym is abusing steroids, but certainly some people are. On the internet these drugs are very easily available, but for non-professional users there is an additional risk. Because most forms of anabolic steroids are not so effective when taken in tablet form (and come with the additional risk of liver disease), they are usually injected. Professional sport teams have access to medical equipment, including high-quality steroid preparations and clean needles, which are not available to the amateur user. So there are risks from not being certain about the quality, purity, or dose of the steroid, and the possibility of blood-borne infections caused by sharing needles.

A worrying recent development is the use of anabolic steroids by teenage boys who are trying to change their body shape. Most people are aware of the increase, in the developed world, in the eating disorder, anorexia nervosa. This serious mental illness mostly affects teenage girls and is associated with the pressure on girls to conform to society's 'ideal' female body shape. School and university counselling services are alert to this significant health problem and offer support and guidance to girls at risk. What most people are not aware of, however, is that teenage boys are subject to very similar pressures to have a 'model' physique. In their case the desired body shape emphasizes not thinness, but muscularity. While anorexia nervosa affects only a small proportion of boys, there is a significantly higher number who turn to anabolic steroids to try to lose fat and gain muscle. Studies in the USA and Australia have found that, for every teenage girl with anorexia nervosa, there is a boy who is abusing anabolic steroids. The boys who are most likely to use steroids are those who feel that they are either too fat or too skinny, and they use anabolic steroids to try to achieve the *toned* look that sports stars and male models display. As we have seen, steroids are readily available on the internet and carry significant risks. In teenagers there is increasing evidence of additional risks to the developing body and brain. Anabolic steroids might appear to be a quick fix to getting a lean muscular look, but the effects are temporary unless the drug use continues and the risks of doing this are high. Unless a teenage boy really wants acne and the risk of serious health problems, a much better bet is to do some exercise; it has the same effect on muscle bulk and body shape as testosterone, without any of the risks.

Growth hormone

The use of growth hormone as a performance-enhancing drug was headline news in the 1998 Olympic Games in Sydney, when a member of the Chinese swimming team was found to have a large amount of the hormone in their luggage; enough, it was suggested, to treat the entire team for the whole championships. Several years later, the team doctor admitted giving athletes a mixture of anabolic steroids and growth hormone to improve their performance. Growth hormone is a large polypeptide hormone which is only active when injected. Like the anabolic steroids, it increases lean body mass by decreasing adipose tissue and increasing muscle. Unlike steroids, however, there is no evidence that it increases muscle strength. As it is very expensive, it has really only been used by professional sports teams under the supervision of corrupt doctors. The adverse effects of using growth hormone are serious, as it increases muscle bulk throughout the body; this includes the heart muscle. There are major heart problems associated with excess growth hormone in the body. In spite of this, the Chinese swimming team clearly thought that it was worth the risk, and the disgrace of being caught and banned.

Erythropoietin (EPO)

In endurance sports, such as long-distance running, swimming, and cycling, we know that increasing the oxygen-carrying capacity of the blood is a good way to improve athletic performance. Athletes can achieve this naturally by going to training camps at high altitude for a couple of months before competing. At high altitudes the concentration of oxygen in the air is lower than at sea level and the body responds by triggering a homeostatic mechanism to increase the number of red blood cells so that the available oxygen can be used most efficiently. When we are put in a low oxygen environment the kidney releases a peptide hormone called erythropoietin, which is also known as EPO. EPO works by stimulating bone marrow to produce more red blood cells, called erythrocytes. With a higher concentration of erythrocytes in blood, the capacity of blood to carry oxygen is increased. This allows the delivery of more oxygen to muscle, enabling the muscle efficiency to increase so the muscle can work harder for longer. This produces an improved athletic performance when the athlete returns to normal altitudes.

Recombinant human EPO was first produced in the 1980s and used to treat patients who have an EPO deficiency. This is usually the result of a severe kidney disease. A certain level of EPO production is necessary to maintain normal levels of erythrocytes in the blood and prevent anaemia. It is a complex hormone to make in the laboratory because it contains a large number of sugars attached to the peptide chain. About 40% of the mass of EPO consists of sugars and 60% amino acids. The half-life in blood of EPO is around five hours, which means that it takes five hours for the blood level to fall by half under normal circumstances. In practical terms this means that EPO needs to be injected more often than once a day. As a result, much of the effort put into developing a synthetic version of the hormone has worked on making it have a longer half-life. The availability of synthetic EPO has made it an attractive drug for some groups of athletes.

Training at high altitudes has always been considered an entirely natural and legitimate way of improving athletic performance, but using injections of EPO is banned. There is no doubt that there are serious health risks in using EPO. Altitude training, as we have seen, triggers a physiological mechanism to raise erythrocyte levels. Injecting EPO can over-stimulate erythrocyte production, far beyond normal physiological levels. This can affect how blood behaves. More erythrocytes in the circulation makes blood thicker and stickier, which makes it more likely to clot. This causes an increased risk of serious health conditions, particularly deep vein thrombosis and pulmonary embolism (blood clots in the circulation to the lung), which can be fatal.

Testing for EPO, like other naturally occurring hormones, is difficult and expensive. At first, anti-doping authorities set a maximum concentration of red blood cells (termed haematocrit) which they argued was the top of the *normal* range. Anything higher than this, they reasoned, was likely to be a result of cheating. It wasn't, however, absolute proof of cheating, so any athlete whose haematocrit was higher than the set level was tested for EPO. As we have seen, EPO is a glycoprotein which contains a very high proportion of sugars. The recombinant versions of EPO which are available to treat patients have significant variations in these sugar residues. In this way it is possible to tell the difference between EPO produced by the kidney and EPO that has been injected. This has enabled anti-doping authorities to develop much more robust tests for EPO use.

Case study 6.2

Hormone abuse in professional cycling

Professional cycling has a long history of drug taking, dating back to the 1880s. In 1930 there was such widespread drug use that a guidance leaflet reminded competitors in the Tour de France that the organizers would not be providing drugs, so they needed to bring their own. At this time most of the drug use involved amphetamines and other stimulants, designed to aid cyclists' endurance. In 1960 a cyclist died at the Olympic Games in Rome, after taking a mixture of amphetamines and alcohol. This led to the introduction of new anti-doping rules, prohibiting the use of any performance-enhancing drugs. However, this didn't prevent the death of a British cyclist, after consuming alcohol and other drugs, in the Tour de France in 1967.

The ban didn't stop the use of drugs and by the late 1970s hormones were included in the mix of performance- enhancing drugs, with corticosteroids and anabolic androgenic steroids the first to be detected. The corticosteroids were used to reduce inflammation and speed up recovery from minor injuries, while the androgenic steroids were used to increase muscle bulk and strength.

Sophisticated testing processes were introduced which could tell the difference between the hormones produced by the body and those taken as drugs. This was done by comparing ratios of different steroid metabolites. Soon the drug users were injecting mixes of steroids to try and mimic the natural profile of metabolites. As the testing was carried out on urine samples, there were also several incidents reported where cyclists had tried to outwit the testers by using the urine from another person!

In the 1990s cyclists began to use EPO to boost their performance. Increasing the number of red blood cells allowed muscles to work harder, especially if the cyclist had previously used steroids to build up muscle bulk. This is a very risky thing to do because of the health dangers. A significant problem in long-distance cycling is dehydration. A cyclist needs to drink large amounts of water during a race to prevent dehydration. If they don't, the chances of a blood clot forming become much higher.

Perhaps the most famous cyclist to be banned for use of performance-enhancing drugs is Lance Armstrong (see Figure A), who won the Tour de France seven times between 1998 and 2005. He first tested positive for a corticosteroid, but was given only a reprimand because he produced evidence that he was using a steroid skin cream for medical reasons. He managed to evade all the drugs tests up until 2012, when he finally admitted using multiple hormones, including EPO, growth hormone, and androgenic steroids to improve his performance. In 2012 Lance Armstrong was publicly disgraced and stripped of all his cycling medals and his Tour de France wins.

Figure A Lance Armstrong—supercheat. The athlete seemed to be a superman, winning the Tour de France seven times. Unfortunately, he was simply a very effective cheat, and his performances were the result of abusing EPO, anabolic steroids, and growth hormone.

LAURENT REBOURS/AP/Shutterstock

It was very clear that the drug cheating in cycling was the work of the whole team, not just the decision of a single athlete. It took an enormous amount of scientific knowledge to give these hormones in a way that boosted performance but could not be detected in a cyclists' urine or blood when they were competing. From the number of cyclists who have been found to have performance-boosting substances in their blood during the period from the late 1990s to the mid 2000s, it appears that about 95% of professional cyclists were cheating.

While EPO was getting the headlines, anabolic androgenic steroids never really went out of favour. In 2006 another Tour de France winner, Floyd Landis, was caught cheating by testers, who used the hormone:metabolite ratio to catch him. The maximum ratio of testosterone to its metabolite, epitestosterone, is 4:1. Usually, it is even lower than this, but a 4:1 ratio is considered to be the upper limit of the normal range. A higher ratio shows that the person has been injecting testosterone. Floyd Landis had a ratio of 11:1. He was disqualified and banned for two years.

Since 2011 and the investigation into the cheating of Lance Armstrong and other professional cyclists, the sport has made a huge effort to clean up its act. Out-of-competition testing has made it more difficult to avoid detection and professional cycling has been put under the microscope, with every medicine used by competitors scrutinized closely. It remains a highly competitive sport, with large financial rewards for winners. The temptation to cheat will never go away, but the very high probability that cheats will be exposed and disgraced appears to be a significant deterrent.

Of course, there can be completely legitimate reasons why an athlete might need to take hormones, particularly if they have an illness which needs treatment with a hormone-based drug. The Olympic rower Sir Steve Redgrave needed to use insulin injections, which would usually be banned, to control his type 1 diabetes. Because of his diabetes he was given medical exemption and allowed to use insulin even when he was competing. This is a really sensible rule, allowing athletes to use *banned substances* when they have a properly certified medical need and is called *therapeutic use-exemption* (*TUE*). The athlete has to be assessed by two doctors, who decide whether the request to use certain drugs is reasonable. There are concerns that the medical exemption rule may be abused. Some cyclists have obtained a TUE to be allowed to use a particular asthma treatment, triamcinolone. This is an anti-inflammatory glucocorticoid which dilates the airways and allows easier breathing. It has been suggested that this is a much stronger form of treatment than these cyclists really need in order to control their asthma and so the cyclists using triamcinolone are gaining an unfair advantage.

❓ Pause for thought

It has been proposed that all therapeutic-use exemptions should be banned. This would mean that athletes with asthma, or with insulin-dependent diabetes would not be allowed to compete. What do you think about this idea? Should there be a sort of Paralympics for athletes who rely on banned substances to keep them healthy?

6.5   Hormones in agriculture: use or abuse?

Most people have a rather romantic view of farming, especially when animals are raised for milk production. We like to think of cows out in a meadow, eating grass, and lying down to chew the cud before gathering twice each day to be led by the farmer into the milking parlour. There is actually a lot of science in farming, which ranges from analysis of grass production to fully automatic milking parlours, including machines which make individual calculations of the amount of supplementary feed each animal needs. In some countries, but not in the UK and the EU, hormones are used routinely to treat cows and to increase both milk and meat production.

Like all mammals, a cow produces milk when she has given birth, in order to feed her calf. As the calf suckles, the cow produces a hormone, prolactin, from the pituitary gland. Prolactin acts on the mammary tissue, called an udder in cows, stimulating the production of milk, as we saw in Chapter 4. She will naturally produce milk until the calf is able to eat grass itself and so no longer depends on its mother's milk. Animal breeders have, over the past 200 years, chosen to breed from those cows that produce the highest amounts of milk. In this way they have developed breeds of cows which are specialized for milk production. A few days after giving birth the calf is usually separated from its mother. A milking machine is used to mimic the effect of the calf suckling and collect the milk. As long as the cow is milked regularly her pituitary gland will continue to produce prolactin and she will continue to produce milk. Typically a calf will drink about ten litres of milk every day (see Figure 6.8). A dairy cow in the UK produces around twice this amount of milk daily.

In the USA, hormones are used to increase the milk yield of dairy cows. Cows are injected with the version of growth hormone found in cows, which is called bovine somatotropin. Growth hormone is very similar to prolactin and its use can increase milk production by as much as 15%. It has been argued that the use of this hormone makes each animal more efficient, causes a lower environmental impact, and allows the consumer access to cheap milk. Opponents of the use of the hormone argue that it

Figure 6.8 An English Longhorn cow suckling her calf. This traditional breed is not usually intensively reared and milked.

Tim Gainey / Alamy Stock Photo

threatens the health and welfare of the cows. There is also an argument that it can have an impact on the health of people who drink the milk, either by causing a change in the milk composition or by transfer of the hormone itself into the milk.

In beef production a different set of hormones is used. In this industry in some countries, calves receive a steroid hormone implant to boost their growth. The implant is a hormone pellet that slowly releases steroids over a period of weeks. The hormones are either oestrogens, androgens, or a combination of both. Studies on weight gain by calves show that the use of oestradiol implants alone causes a 14% increase in daily weight gain compared with untreated calves. When the hormone is combined with other growth promoting drugs the increased weight gain is 41%. The meat that comes from these cows has little fat and so is very marketable, because lean meat is what many consumers want. There are widespread fears about the synthetic hormones contaminating the meat and affecting human health. Because of this the EU has banned all hormone-treated meat, including beef, from its markets.

It has been suggested that the human digestive system is quite capable of breaking down any hormones that come from meat and dairy products. It is argued that the animals producing the meat and milk will produce their own hormone so we are exposed to these without harm to ourselves. It is worth bearing in mind that the steroids used as growth promoters are not the same chemicals as naturally occurring hormones. They are all synthetic variants of natural steroids, which can behave very differently when we ingest them. It is also now known that not all proteins are broken down into their individual amino acids before absorption from the gut. In fact, proteins can cross from the gastrointestinal tract into our blood intact. The smaller the protein or peptide, the more likely this is to happen. This is especially true in babies and small children and in adults with disorders of the gut lining. The levels of hormone used in intensive agriculture are also significantly higher than the levels found naturally in the animals. The other issue, of course, is about animal welfare and the lengths to which we should go in manipulating animals to provide the cheap and plentiful food we want.

In this chapter we have seen how hormones are used to treat both endocrine and inflammatory diseases and to improve symptoms of menopause. We have considered the methods originally used to produce hormones for medical use and seen how technological advances have allowed the production of safer and more pure preparations of hormones. We have reviewed instances of hormone abuse, considering both thyroid hormone and a group of performance-enhancing hormones. Finally, we reviewed the use of hormones in animal husbandry, considering both sides of the debate. All of these areas raise many questions—and the answers to at least some of them are way beyond the remit of scientists. Questions about cheating in sport or acceptable welfare standards for animals are questions for society to answer—science can only provide the evidence of risk and benefit on which we can base those choices.

Chapter summary

- When the body is not making enough of a hormone to keep healthy it is usually necessary to give the hormone as a prescribed medicine. Steroid hormones and thyroxine are usually given as tablets, taken orally, but peptide hormones usually need to be given by injection.

- The term hormone replacement therapy (HRT) has come to mean the replacement of oestrogens in post-menopausal women, given as a daily tablet containing both oestrogen and progesterone.

- Peptide hormones used for replacement therapy were originally obtained from animals. Recombinant technology has allowed the synthesis of human peptides in large quantities from microorganisms.

- In spite of significant adverse effects, synthetic corticosteroids are widely used to treat inflammatory illnesses and asthma.

- Anabolic androgenic steroids, corticosteroids, growth hormone, and erythropoietin have all been systematically abused throughout sport for decades. Although each has been identified with performance benefits, they also cause harm.

- The use of hormones in agriculture remains controversial. There is no doubt that productivity can be increased but many people have concerns about both animal and human welfare issues as a result.

Further reading:

https://www.fda.gov/AnimalVeterinary/SafetyHealth/ProductSafetyInformation/ucm055435.htm
This is the link to the USA government website information about the use of bovine somatotropin in milk production.

https://web.archive.org/web/20080904003449/http://ec.europa.eu/food/fs/sc/scah/out21_en.pdf
And this is the link to the EU's report on the impact of using bovine somatotropin on animal health and welfare.

http://extension.uga.edu/publications/detail.html?number=B1302#Stocker
This is a publication from the University of Georgia (USA) promoting the use of growth promoters in beef production.

http://ldb.org/menopaus.htm
This is an interesting scholarly article on the menopause.

Patisual, H.B. and Jefferson Front, W. (2010). 'The pros and cons of phytoestrogens', *Neuroendocrinol.* 31(4), 400–19. doi:10.1016/j.yfrne.2010.03.003.
An interesting and very balanced review of the pros and cons of plant oestrogens.

Benedek, T.G. (2011). 'History of the development of corticosteroid therapy', *Clin. Exp. Rheumatol.* 29(68), S5–S12.
For a more detailed account of the use of corticosteroids as anti-inflammatory agents look at this free review article.

http://joe.endocrinology-journals.org/content/170/1/27.full.pdf
This review article summarizes what we know about the effects of testosterone on muscle.

≋ Discussion questions

6.1 Should we allow athletes to use whatever they like to improve their performance, or do we have a duty of care to try to prevent people from harming themselves for sporting benefit?

6.2 Do you feel well enough informed to have an opinion about the use of hormones in farm animals? What more would you like to find out?

6.3 Do the oestrogens found in *natural* remedies, such as soy products, act in a different way to conventional HRT? Why do you think that some people prefer to use soy products rather than take conventional HRT?

GLOSSARY

Addison's disease A form of adrenal hormone deficiency, caused by autoimmune disease.

Adenoma A non-malignant growth of hormone-secreting cells, resulting in an excess of the hormone.

Adrenal cortex The outer part of the adrenal gland, producing steroid hormones. Major hormones produced are cortisol, aldosterone, and DHEA.

Adrenal medulla The inner part of the adrenal gland producing the catecholamines, adrenaline, and noradrenaline. It is a modified neural ganglion, part of the sympathetic nervous system.

Adrenaline A catecholamine hormone derived from the amino acid tyrosine. Released by the adrenal medulla and sympathetic nerve endings. Causes fight-or-flight response.

Adrenocorticotropic hormone (ACTH) A medium-sized peptide hormone from the pituitary gland. Controls the production of cortisol (a glucocorticoid) by the adrenal cortex.

Andiostenedione Androgen found in the blood of male and female spotted hyenas.

Antidiuretic hormone The old name for vasopressin.

Antisense RNA A single-stranded RNA that is complementary to a specific mRNA with which it hybridizes, and thereby blocks its translation into protein.

Apoptosis Programmed cell death.

Arcuate nucleus The region of the hypothalamus involved in the control of appetite.

Autistic spectrum disorder A group of developmental disorders that affect a person's social interaction, communication, behaviour, and interests.

Autoantibody An antibody produced by an individual animal in an abnormal response to a constituent of its own tissues.

Autoimmune disease A disease caused by the production of autoantibodies.

B

Bioassay A method for measuring hormone levels based on the biological action of the hormone.

Biologically active Used to refer to the fraction of hormone that is capable of binding to and activating a receptor.

C

Catecholamines A group of hormones produced from the amino acid tyrosine. Includes dopamine, adrenaline, and noradrenaline.

Chaperone proteins In endocrinology **chaperones** are **proteins** that assist assembly or disassembly of hormone receptors and the passage of proteins through the cell.

Cholesterol An important component of cell membranes and a precursor for steroid hormone synthesis.

Circadian rhythm Any biological process that has a predictable cycle of activity over twenty-four hours.

Congenital adrenal hyperplasia (CAH) CAH is a form of adrenal insufficiency in which the enzyme needed for producing two important adrenal steroid hormones, cortisol and aldosterone, is deficient. Because cortisol production is impeded, the adrenal gland instead overproduces androgens.

Cryptorchidism Literally 'hidden testis'. A condition seen in baby boys where one of the testes has not descended into the scrotum.

Cyclic AMP Cyclic form of adenosine monophosphate; it acts as an intracellular amplifier or second messenger of signals derived from hormones or neurotransmitters.

D

Dalton A standard unit of mass, used in chemistry and biochemistry, that quantifies mass on a molecular scale (molecular mass).

Diabetes insipidus Disorder where there is excess production of urine as a result of either lack of vasopressin (ADH) production or a receptor deficiency.

Dihydrotestosterone (DHT) The activated form of testosterone.

Dimer (receptor) A unit formed by the joining of two molecules (receptors).

Diurnal rhythm See circadian rhythm

Dopamine A catecholamine best known as a neurotransmitter, which also acts as a hormone to regulate prolactin secretion.

Dose dependent A term used to describe an effect where the size of the effect is directly related to the amount of hormone used; more hormone gives a bigger effect.

E

Ecdysteroids Insect steroid hormones involved in reproduction and moulting.

Ectopic hormone production The production of a hormone in a different organ tissue from normal, caused by expression of a gene in a non-endocrine tissue.

Endocrine system The collection of glands, tissues, and cells that produce and respond to hormones, including the hormones themselves.

Endocrinology The study of hormones.

Endocytosis A form of transport in which a cell transports molecules into the cell by engulfing them.

Endoplasmic reticulum Intracellular organelle; can be rough or smooth. Rough ER contains ribosomes and is important in peptide synthesis. Smooth ER does not have ribosomes and is important in steroid synthesis.

Enzyme-linked immunosorbent assay (ELISA) A test that uses antibodies and a colour change reaction to measure substances in biological samples.

Epinephrine Another name for adrenaline, one of the catecholamines.

Exocrine pancreas The part of the pancreas that produces digestive enzymes, etc., not hormones.

F

Follicle stimulating hormone (FSH) Large glycoprotein hormone produced by the pituitary gland. Important in gamete production.

G

Ghrelin Hormone released by the stomach which acts on the brain to increase a feeling of hunger.

Glucagon Pancreatic hormone which is released in response to low blood glucose and acts to increase blood glucose levels.

Glucocorticoid The most important type of steroid hormone produced by the adrenal cortex. The main glucocorticoid is cortisol.

Goitre A swelling in the front of the neck caused by an enlarged thyroid gland. Goitre can be a result of either overactivity or underactivity of the thyroid gland.

Gonad The main reproductive gland, making hormones and gametes. The male gonad is called a testis (plural testes) and the female gonad is an ovary.

H

Half-life The half-life of a hormone usually refers to the length of time it takes for half the amount of the hormone present in blood to be excreted or metabolized.

Hormone synthesis Production of the hormone.

Human chorionic gonadotropin (hCG) A large glycoprotein hormone, produced throughout pregnancy. Measurment of this hormone is the basis of the home pregnancy test.

Hyper- Refers to an excess of a hormone; so hyperthyroidism is an excess of thyroid hormone.

Hypo- Refers to a deficiency of a hormone; so hypothyroidism is a lack of thyroid hormone.

Hypospadia A developmental abnormality of the penis in which the urethra opens along the shaft of the penis, instead of at the tip.

Hypothesis A suggested explanation for an observed phenomenon. A scientific hypothesis must be proposed in a way that means it can be tested.

I

Immunoassay A biochemical test that uses antibodies to measure the concentration of a biological substance.

In vitro method Literally 'in glass'. A biological process or investigation taking place outside a living organism, for example in a test tube or petri dish.

Intracellular signalling pathway The mechanism for taking a hormone signal, detected by a cell-surface receptor, into the cell so that the cellular function is altered.

J

Juvenile hormones Insect hormones that control the stage of development of the instar at each moult.

K

Knock-out gene technology A genetic technique where one of an organism's genes is made to be inactive: 'knocked out'.

L

Leptin A peptide hormone produced by adipose tissue which acts on the brain to suppress hunger and eating.

M

Melanocyte stimulating hormone (MSH) A peptide hormone from the pituitary, closely related to adrenocorticotropin (ACTH).

Menarche The first menstrual period in the reproductive life of a woman.

Menopause The end of a woman's reproductive life, when she stops having periods and is no longer able to conceive.

Mitochondria Rod-like structures with inner and outer membranes that are teh site of aerobic respiration in the cell.

Monoclonal antibody A type of antibody production that involves cloning of the antibody-producing cell, so that each cell produces identical antibodies.

Monogamous Having only one sexual partner, sometimes called *mating for life*.

N

Negative feedback control A control mechanism where the product of a chain of reactions (or a hormone cascade) acts back on the early steps to suppress further production.

Neurohormone A hormone produced by nerve cells and released into blood from nerve terminals.

Neuropeptides Signalling molecules in insects and many other multicellular organisms.

Noradrenaline A catecholamine hormone produced by the adrenal medulla and involved in the fight-or-flight response.

Norepinephrine Another name for noradrenaline.

O

Oestrogens A group of steroid hormones produced by the ovary. The female sex steroids.

Osteoporosis Loss of bone mass often seen in post-menopausal women, associated with increased risk of broken bones.

Ovary The female gonad. Makes female sex steroids and produces ova (eggs).

Ovulation The release of a mature ovum (egg) from the ovary.

Oxytocin Small peptide hormone released from the pituitary gland. Has a role in reproduction and more widely in behaviour.

P

Pheromone A chemical substance produced and released into the environment by an animal, especially a mammal or an insect, affecting the behaviour or physiology of others of its species.

Phytoestrogens A group of oestrogen-like steroids made by plants.

Precursor peptide A peptide before it reaches its final state.

Precursors Chemicals that come before others in a pathway of synthesis.

Pre-hormone A hormone that is not fully activated.

Prostaglandin A group of physiologically active lipids that have hormone-like effects, usually quite local to the site of production.

Protein kinase An enzyme that modifies other proteins by adding a phosphate group to them.

R

Rate-limiting step The slowest step of a process, which determines the overall rate at which that process can happen.

Receptor agonist A substance that binds to the same receptor as a hormone, and activates the receptor in the same way as the hormone does.